MANUFACTURING TECHNOLOGY

M. HASLEHURST
D.I.C., C.Eng., M.I.Mech.E., M.I.Prod.E.

Dean of the Faculty of Technology and Science,
North Cheshire College

HODDER AND STOUGHTON

LONDON   SYDNEY   AUCKLAND   TORONTO

Higher Technician Series

General Editor
M. G. Page,
B.Sc., C.Eng., M.I.Mech.E., M.I.Prod.E., M.B.I.M., F.S.S
Head of Department of Production Engineering,
The Polytechnic, Wolverhampton

British Library Cataloguing in Publication Data

Haslehurst, M.
   Manufacturing technology. – 3rd ed.
   1. Machine-shop practice
   I. Title
   670.42'3   TJ1160

   ISBN 0 340 26980 4

First printed 1969
Reprinted 1970
Second edition 1972
Reprinted 1974, 1975, 1977 (with modifications), 1978, 1979, 1980
Third edition 1981

Printed and bound in Great Britain
for Hodder and Stoughton Educational
a division of Hodder and Stoughton Limited
by Hazell Watson & Viney Ltd, Aylesbury, Bucks

Preface

MOST HIGHER National Certificate in Engineering Courses, and CGLI Technician courses have been phased out to be replaced by appropriate Technician Education Council (TEC) programmes. Many Technician Higher Certificate (HC) programmes approved by the TEC A5 Programme Committee (Mechanical and Production Engineering) include the following TEC units of study:

TEC U77/423 Manufacturing Technology IVA
TEC U79/608 Manufacturing Technology V
College devised Manufacturing Technology IVB

In 1979 my text book *Technician Manufacturing Technology IV* was published which specifically covered the learning objectives set out in TEC unit U77/423. It seemed sensible that I should revise my other publication *Manufacturing Technology* in order to accommodate much of the latter two units listed above. I have attempted to do this, bearing in mind that the content of the college devised unit IVB will obviously vary from college to college.

This edition therefore is updated and revised with the above objectives in mind. I hope that it will prove to be a valuable learning aid to students studying the level IVB and U79/608 level V TEC units respectively. In addition I have endeavoured to retain the essential nature of the book in order that it may continue to serve some useful purpose for Higher National Diploma and Degree students.

M.H., 1981

Contents

CHAPTER 1

Manufacturing Costs

1.1 AVAILABILITY OF ACCURATE COSTS

IN ORDER for a manufacturing organisation to remain competitive it must make its products at the minimum cost consistent with the required quality and function of the product. The company will then make the maximum profit possible which will ensure the continuing health of the organisation. (We will assume for the purposes of this book that minimum costs will give maximum profits although an economist will assert that this is not always so.) A production engineer with a good knowledge of manufacturing processes must always be cost conscious, and should use information on costs to decide which of two or more perfectly feasible processes should be used to manufacture a particular product.

If costs are to be used in this way then it is essential that accurate costs are made available to the production engineer. This is possible if a company has a cost control system in which actual costs are constantly being compared to estimated or planned costs to give a measure of the efficiency of the manufacturing unit. Information made available to the production engineer for analysis must not be historical and hence largely worthless. The introduction of cost accounting systems, like standard costing, in the last few years, has proved to be of immense value both to accountants and engineers. A standard costing system allows the technique of variance analysis to be used in which actual and standard costs are compared. The analysis of variations between the costs often provides vital cost control information.

1.2 TYPES OF COSTS

Costs of engineering products can broadly be grouped under *direct costs* or *indirect costs*. *Direct costs* are the costs of those factors which can be directly attributed to the manufacture of a specific product. These are the costs of *material* and *labour*. *Material cost* is the cost of that material which goes into the finished product and includes all waste which has been cut away from the original bar, casting etc. *Labour cost* will be the product of the number of pieces produced and the piecework rate (in the case of a simple incentive scheme) or the product of the time spent in manufacturing the product by the direct shop floor workers and the wage rate (in the case of standard times systems). In the case of more complicated wage schemes for direct workers such as premium bonus systems the calculation of direct labour costs will be more difficult. *Indirect costs* are the costs of those factors which can only be indirectly attributed to the manufacture of a specific product. They are sometimes called *overheads* or *oncosts*. They can be subdivided for convenience under three headings:

a) *Works overheads.* These consist of the cost of the wages of works superintendants, foremen, inspectors, storekeepers, labourers etc., cost of cutting oil, depreciation of machines, heating, lighting, rents, rates, etc.

b) *Office overheads.* These consist of the cost of the wages of all office staff, postage, legal expenses, depreciation of office equipment, etc.

c) *Sales overheads.* These consist of the cost of the wages of all sales staff, advertising, sales commissions, etc.

Therefore it can be seen that indirect costs are the total costs of running the organisation less the direct material costs and the direct labour costs. The major difficulty in dealing with indirect costs is to decide accurately how much of the total overheads should be borne by a particular component or batch of components. It is obvious that if this is not done reasonably accurately then the sales price for the product will be unrealistic, where sales price = direct costs + overheads + profit.

Overheads are based upon past experience and in much engineering work are expressed as a percentage of the direct labour cost. If the total cost of manufacturing and selling last year was £850 000, and in that time the cost of direct labour was £200 000, while the cost of direct material was £50 000, then the overheads would be £600 000. Past experience in this case shows that overheads are 300% of direct labour costs. This value could be used until more up to date costing information proves it to be inaccurate, and a better estimate made. Any efficient manager is aware of the effect of over-large indirect costs on the prosperity of a company, and is always seeking ways and means of reducing them.

The total cost of a product is the direct cost of manufacturing the product plus any indirect costs attributed to the manufacture of the product.

Example 1.1

A batch of 500 components is produced on a capstan lathe. The piece work rate/piece is 2·5p, and the direct material cost/piece is 4p. Overheads are 450% of direct labour cost. What is the total cost of the batch of components?

Solution

$$\text{Direct material cost} = 500 \times 4\text{p} = 2\,000\text{p}$$
$$\text{Direct labour cost} = 500 \times 2\cdot5\text{p} = 1\,250\text{p}$$
$$\text{Indirect cost} = 1\,250 \times \frac{450}{100} = 5\,625\text{p}$$

$$\text{Total cost} = 8\,875\text{p}$$
$$= £88\cdot75$$

In order for the production engineer to use cost data as a tool to help analyse manufacturing problems, costs may be more conveniently grouped under *fixed costs* and *variable costs*. It will also be found that this grouping often fits better into a standard costing system.

Fixed costs are those costs which are independent of the quantity of the product manufactured. These include preparation costs such as the cost of tooling, setting-up, etc., and also the interest costs and depreciation costs. Fixed costs include all those expenses which must be borne by the firm whether one or a thousand of the products are made. Obviously the more products that are made, the less will be the fixed cost/piece.

Variable costs are those costs which vary as the quantity of products made varies. Usually variable costs increase proportionally as the number of products made increases, and includes the direct labour and material costs, and also that part of the indirect costs which will vary as production varies. The total cost of a product then can also be seen to be fixed cost + variable cost.

Example 1.2
Segregate the fixed cost and the variable cost of running a motor car for one year.

Solution
Fixed cost (independent of distance covered in the year)
i) Depreciation/year — this is the amount by which the car depreciates in value each year.
ii) Interest costs/year — this is the amount of interest the capital spent on the car would have earned if the capital had been invested.
iii) Road tax and insurance.
iv) Garage rent (if any).
v) Maintenance costs — that proportion of total maintenance costs which would be incurred even if the car never left the garage.
Variable cost (varies proportionally to distance covered in the year)
i) Cost of petrol.
ii) Cost of oil.
iii) Maintenance costs — that proportion of total maintenance costs which would be incurred due to driving the car, i.e. the greater part of the maintenance costs.

Obviously the greater the distance covered in the year, the less will be the fixed cost/kilometre, and the less will be the total cost/kilometre. This is the essence of the philosophy of production engineering, viz. the longer the production run the better.

It can be seen from the above in the case of maintenance costs that the precise allocation of costs into a fixed or a variable category is not always easy. However, if the accountants achieve this accurately and make the information available to the engineer, he has the means of comparing processes by using what are known as *break-even charts*.

1.3 BREAK-EVEN CHARTS

These comparatively simple charts give the production engineer a powerful tool by which to compare feasible alternative processes. The fixed and variable costs for two or more alternative processes are plotted on a graph to some suitable scale as shown in Fig 1.1.

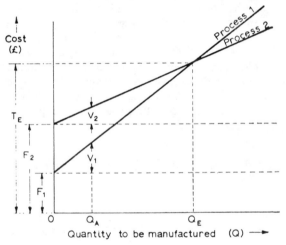

Fig 1.1 Break-even Chart for two processes.

F_1 = Fixed costs for process (1)
F_2 = Fixed costs for process (2)
V_1 = Variable costs for process (1)⎫
V_2 = Variable costs for process (2)⎬ at quantity Q_A
Q_E = Break-even quantity
T_E = Total costs of manufacture at quantity Q_E.

The reasonable assumption is made for each process that the variable cost is a linear function of the quantity manufactured. Therefore, once the fixed costs have been plotted, only one value for the variable costs is required at some value Q_A and the total cost lines can be drawn.

Where these lines intersect is known as the break-even point, i.e., the point where the total cost of manufacture of quantity Q_E is identical for both process (1) and process (2). The break-even chart tells us to:

use process (1) if the quantity to be manufactured $\leqslant Q_E$

use process (2) if the quantity to be manufactured $\geqslant Q_E$

The value of Q_E can be scaled directly from the chart with sufficient accuracy, although it can easily be calculated if preferred.

Example 1.3

A component can be produced with equal facility on either a capstan lathe or a single-spindle automatic. Find the break-even quantity Q_E if the following information is known.

	Capstan Lathe	Automatic
a) *Tooling Cost*	£30·00	£30·00
b) *Cost of Cams*	—	£150·00
c) *Material Cost/Component*	£0·25	£0·25
d) *Operating Labour Cost*	£2·50/h	£1·00/h
e) *Cycle Time/Component*	5 min	1 min
f) *Setting up Labour Cost*	£4·00/h	£4·00/h
g) *Setting up Time*	1h	8h
h) *Machine Overheads*		
(setting and operating)	300% of (d)	1 000% of (d)

Solution

Capstan Lathe Overheads $= \dfrac{300}{100} \times 2\cdot50 = £7\cdot50/h$

Fixed costs $=$ tooling cost $+$ setting-up cost
$= 30\cdot00 + 1(4\cdot00 + 7\cdot50)$
$= 30\cdot00 + 11\cdot50 = £41\cdot50$

Variable costs/component $= (2\cdot50 \times \tfrac{5}{60}) + 0\cdot25 + (7\cdot50 \times \tfrac{5}{60})$
$= 0\cdot21 + 0\cdot25 + 0\cdot63 = £1\cdot09$

Variable costs/1 000 components $= £1\,090\cdot00$

Automatic Overheads $= \dfrac{1000}{100} \times 1\cdot00 = £10\cdot00/h$

Fixed costs $=$ tooling cost $+$ cam cost $+$ setting-up cost
$= 30\cdot00 + 150\cdot00 + 8(4\cdot00 + 10\cdot00)$
$= 180\cdot00 + 112\cdot00 = £292\cdot00$

Variable costs/component $= (1\cdot00 \times \tfrac{1}{60}) + 0\cdot25 + (10\cdot00 \times \tfrac{1}{60})$
$= 0\cdot02 + 0\cdot25 + 0\cdot17 = £0\cdot44$

Variable costs/1 000 components $= £440\cdot00$

These costs can now be plotted on a break-even chart (Fig. 1.2) to find the value of Q_E.

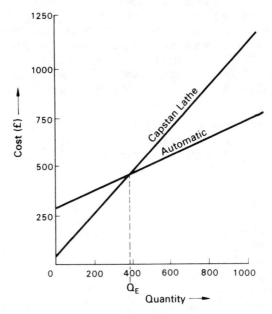

Fig 1.2 Break-even Chart.

Q_E is scaled from the break-even chart (Fig 1.2) and found to be 385
    If the batch size to be manufactured is equal to or less than 385 use the capstan lathe.
    If the batch size to be manufactured is equal to or greater than 385 use the automatic.
    In this example no account was taken directly of the costs of depreciation and interest charges. The accountant's methods of dealing with these charges will be determined by circumstances, and a text book on cost accounting should be consulted if detailed information is required of costing procedures.
    If preferred, the break-even quantity Q_E can be calculated thus:

Total cost of producing a quantity x on the capstan lathe

$$= 41{\cdot}50 + 1{\cdot}09x$$

Total cost of producing a quantity x on the automatic lathe

$$= 292{\cdot}00 + 0{\cdot}44x$$

When x is equal to Q_E the total costs for each machine must be equal.

$$\therefore\ 41{\cdot}50 + 1{\cdot}09x = 292{\cdot}00 + 0{\cdot}44x$$

$$1 \cdot 09x - 0 \cdot 44x = 292 \cdot 00 - 41 \cdot 50$$

$$\therefore 0 \cdot 65x = 250 \cdot 50$$

$$x = \frac{250 \cdot 50}{0 \cdot 65} = 385 \text{ at the break-even point.}$$

Let us now consider a case where the fixed costs of a tool may re-occur after fairly short production runs. A typical example of this occurs in press tool work. Consider the use of a reinforced plastic forming tool (the plastic former usually being strengthed by the addition of glassfibre) which is suitable for short runs, compared to the conventional 'all-metal' forming tool with the former made from hardened and tempered tool steel, more suitable for long runs.

Example 1.4
A blanked component can be formed with equal facility upon a plastic forming tool or a steel forming tool. Find the break-even quantity Q_E if the following information is known.

	Plastic tool	*Steel tool*
a) *Fixed costs*	£700·00	£1 500·00
b) *Variable costs/component*	£0·04	£0·04
c) *Life of former*	9 000	60 000
	components	components

The cost of a replacement plastic former is £450·00. This of course is an additional fixed cost on the plastic tool after a production run of 9 000 components.

Solution
Variable costs/10 000 components for each tool

$$= 10\ 000 \times 0 \cdot 04 = £400 \cdot 00$$

The costs can now be plotted on a break-even chart Fig 1.3 to find the value of Q_E.

Break-even quantity $Q_E = 18\ 000$ components.

Finally, let us consider the comparison of processes which are flexible enough to undertake 'one-off type work', or long production runs, the fixed and variable costs altering accordingly. A good example of this is a numerically controlled milling machine compared to an operator controlled milling machine. With the former machine, once the tape is produced it can be used for one or very large numbers of components. With the latter machine no special tooling is usually required for one-off jobs, but a fixture will be required for the production of batches.

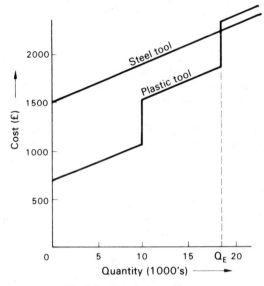

Fig 1.3   Break-even Chart.

Example 1.5
A component can be produced with equal facility upon either a numerically controlled milling machine, or an operator controlled milling machine by what is still a more conventional method.
a) Which process should be chosen for minimum costs if two components only are required? (Assume no special tooling for the conventional machine and hence no fixed costs, but that it is a toolroom universal milling machine with high overheads.)
b) Which process should be chosen for minimum costs if a batch of 100 components is required? (Assume special tooling such as a fixture and gauges are required for the conventional machine, which is a plain miller on a production line.)

The following cost information is known.

	(a)		(b)	
	N/c machine	*Conventional machine*	*N/c machine*	*Conventional machine*
Fixed cost	£80	—	£80	£300
Labour/part	£1·50	£12·50	£1·50	£0·35
Material/part	£1·00	£1·00	£1·00	£1·00
Overheads/part	£3·00	£30·00	£3·00	£2·25

Solution

a) Variable costs/part for N/c machine = £1·50 + £1·00 + £3·00 = £5·50.
Variable costs/part for conventional machine = £12·50 + £1·00 + £30·00 = £43·50.

The costs can now be plotted on a break-even chart (Fig 1.4).

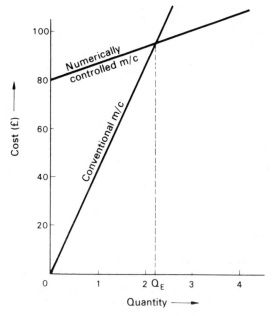

Fig 1.4 Break-even Chart.

Q_E is equal to approximately 2·2 components.

The conventional machine should be chosen if two components only are required.

b) Variable costs/part for N/c machine = £5·50, as before.

Variable costs/100 parts for N/c machine = £550·00

Variable costs/part for conventional machine = £0·35 + £1·00 + £2·25 = £3·60

Variable costs/100 parts for conventional machine = £360·00

The costs can now be plotted on a break-even chart (Fig 1.5) to find the value of Q_E.

Q_E is equal to approximately 116 components.

The numerically controlled milling machine should be chosen if 100 components only are required.

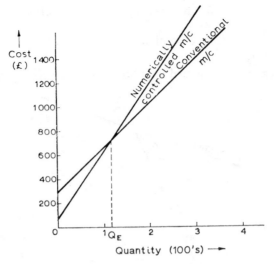

Fig 1.5   Break-even Chart.

1.4 RATIO ANALYSIS

Cost control can be aided by the use of a technique called *ratio analysis*.
Many financial ratios may be used by management, and these can be
displayed in a hierarchy known as a ratio pyramid. The most important
ratio used to analyse financial performance is Profit to Capital Employed and is therefore at the apex of the pyramid. This ratio is a
measure of the rate of return on capital employed (ROCE), and is given
by the formula:

$$\frac{\text{Net profit}}{\text{Capital employed}} \times 100$$

This *primary ratio* is made up of two *secondary ratios* (which form the
second level of the pyramid) as follows:

$$\frac{\text{Net profit}}{\text{Capital employed}} = \frac{\text{Net profit}}{\text{Sales income}} \times \frac{\text{Sales income}}{\text{Capital employed}}$$

Likewise the secondary ratios may be developed to produce *tertiary
ratios* which make up the subsequent levels of the pyramid. This is
shown in simplified form in Fig 1.6 in order to illustrate the principle.

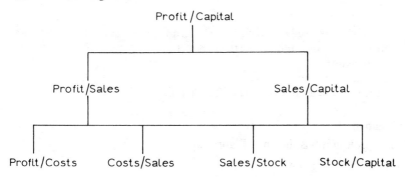

Fig 1.6 Ratio Pyramid.

Profit – Volume Ratio

The secondary ratio in the pyramid which connects profit and sales income (volume) is called a P/V (Profit/Volume) ratio and is useful for comparing the profitability of a product manufactured by different systems or processes. The elements contained in this ratio may best be analysed using a form of break-even chart called a *profit-volume chart* or a *profitgraph*, as shown in Fig 1.7.

The profitgraph shown in Fig 1.7 is a break-even chart in which the fixed costs (F) are indicated as a negative quantity on the vertical ordinate. The break-even quantity (Q_E), above which a profit is made, is given by the intersection of the sloping income line with the horizontal ordinate.

The profitability of the product is given by the slope of the income line. The Profit-Volume ratio (r) depends upon the slope, therefore:

$$r = \frac{F}{Q_E}$$

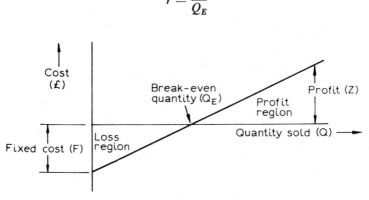

Fig 1.7 Profitgraph.

Example 1.6

A product has fixed costs of £12 000 and a break-even quantity of 17 000 units. Draw a profitgraph for the product.

Calculate:

a) the profit-volume ratio

b) the profit at a sales volume of 24 400 units.

Solution

The profitgraph is shown at Figure 1.8

a) $$r = \frac{F}{Q_E} = \frac{12\,000}{17\,000} = 0.7059$$

b) From the profitgraph it can also be seen that:

$$r = \frac{F + Z}{24\,400}$$

$$\therefore \qquad 0.7059 = \frac{12\,000 + Z}{24\,400}$$

$$\therefore 0.7059 \times 24\,400 = 12\,000 + Z$$
$$\therefore \qquad Z = 17\,224 - 12\,000 = £5224$$

This solution can also be found, with some loss of accuracy, by scaling

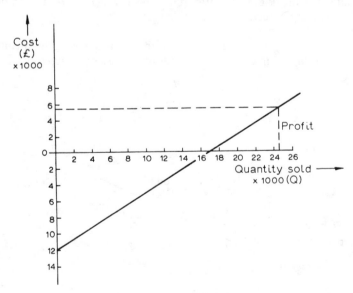

Fig 1.8 Profitgraph.

directly from the profitgraph. Further, the product quantity can be measured either in sales volume (as shown in Fig. 1.8) or in sales income. In the former case the P/V ratio units will be pounds/unit, and in the latter pounds/pound.

Multi-product Profitgraph

The use of the profitgraph may be extended to more than one product or system as shown in Fig 1.9.

Three product costs for products A, B and C are depicted in Fig 1.9. *A* and *C* return a profit and B a loss. The slope of the equivalent income line (shown dotted) can be used to calculate an equivalent P/V ratio for the system. Remember that the slope of the equivalent income line indicates the profitability of the system, which in turn is affected by the profitability of the individual products within the system.

1.5 MARGINAL COSTING

The concept of fixed and variable costs was introduced in Section 1.2. The segregation of costs in this manner has allowed the useful technique of *marginal costing* to be used as a management aid to cost analysis. Marginal costing may be defined as the technique of ascertaining marginal costs, and the effect upon profit of changes in volume by differentiating between fixed and variable costs.

A marginal cost (M) is the amount at any given volume of output by which the total cost (Y) is changed if the volume of output (Q) is changed by one unit; i.e.,

$$\text{Marginal Cost } (M) = Y_Q - Y_{(Q-1)}$$
$$\text{where total cost } Y = (\text{fixed cost}) + (\text{variable costs})$$

The technique can perhaps best be illustrated using a break-even chart similar to that shown at Fig 1.1. In this case the total sales revenue will be compared to the total cost of the product. Consider the following example.

Example 1.7

A product has a selling price of £10.50, and fixed costs of £23 000 have been incurred for its manufacture. The variable cost per product is known to be £4.60. Draw a break-even chart, and determine:

a) the break-even quantity (Q_E) beyond which a profit is made
b) the profit (Z) for a sales volume of 9 500 products
c) the marginal cost (M) at quantity 3 000 products.

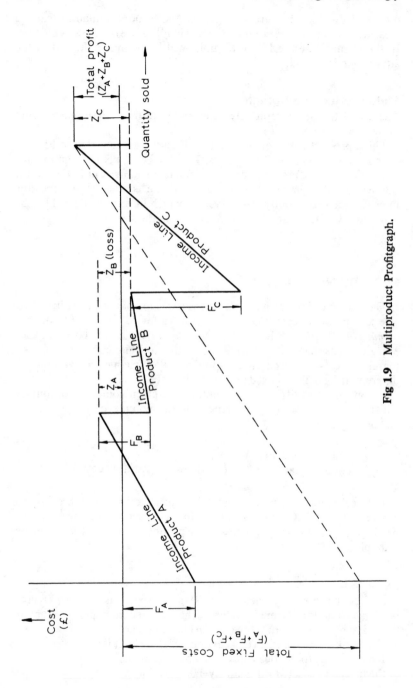

Fig 1.9   Multiproduct Profitgraph.

Solution

At quantity 0 the total sales revenue = 0
At ,, 9 500 ,, ,, ,, ,, = 9 500 × £10.50
= £99 750

∴ the total sales revenue line is a straight line drawn through the ordinates (0,0) and (9 500, 99 750).

At quantity 0 the total cost = £23 000
At ,, 9 500 ,, ,, ,, = fixed cost + variable cost
= £23 000 + (9 500 × £4.60)
= £66 700

∴ the total cost line is a straight line drawn through the ordinates (0,23 000) and (9 500, 66 700).

The break-even chart is shown at Fig 1.10.

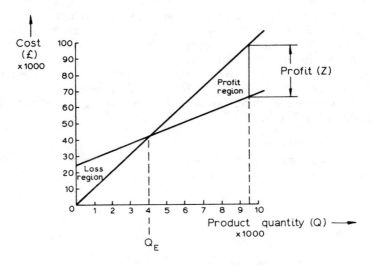

Fig 1.10 Break-even Chart.

a) From Fig 1.10, Q_E is found to be 4 000.
Or by calculation:

At quantity Q_E, $0 + 10.50Q_E = 23\ 000 + 4.60Q_E$
∴ $5.90Q_E = 23\ 000$
∴ $Q_E = 3898$

b) From Fig 1.10, Z is found to be £33 000.
   Or by calculation:

$$Z = \text{Total sales revenue} - \text{Total cost}$$
$$= £99\ 750 - £66\ 700$$
$$= £33\ 050$$

c) At quantity $Q = 3\ 000$:

$$M = Y_Q - Y_{(Q-1)}$$
$$= [23\ 000 + (3\ 000 \times 4.6)] - [23\ 000 + (2\ 999 \times 4.6)]$$
$$= £36\ 800 - £36\ 795.40$$
$$= £4.60$$

This simple example highlights some of the principles of the technique of marginal costing. Here it has been assumed that the variable cost factor is a linear function, hence the marginal cost [as shown at c)] is equal to the variable cost in the above example. This is not always so in practice, in which case the marginal cost would be different at different quantities.

The margin between sales revenue and variable cost represents the contribution of the revenue to fixed cost and profit. Each product sold contributes £5.90 (i.e., £10.50 − £4.60) so that beyond the break-even point of 3 898 each further product sold increases the profit by £5.90. The marginal cost is the minimum value at which the product selling price must be fixed to recover all the costs. Hence, marginal costing provides management with useful information for price fixing.

For interest the cost information depicted in Example 1.7 is shown on a profitgraph in Fig 1.11.

Machine Replacement

The marginal costing principles outlined above may also be used in other areas, such as determining the optimum replacement period for a

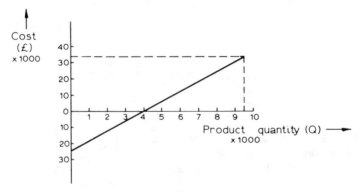

Fig 1.11 Profitgraph.

piece of capital equipment which is deteriorating in value. Here, the total machine cost consisting of the fixed and variable (running) costs is depicted over 'x' years, as is also the machine value. Fig 1.12 illustrates the principle.

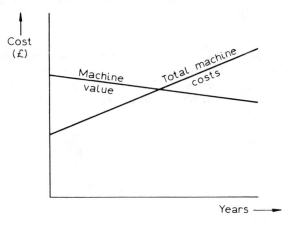

Fig 1.12   Machine Costs.

Fig 1.12 shows that after a period of time the value of the equipment deteriorates below the total cost, this giving a guide to the optimum replacement period. The fixed and variable costs are segregated in a similar manner to that shown for a motor car in Example 1.2. In practice this is a more complex situation than that described, particularly with large capital items, and factors such as taxation, inflation trends, etc., must be taken into account. Further, considerations affecting machine replacement decisions are not only economic, but may involve such matters as availability of specialist labour, material utilization, machine efficiency, etc.

1.6 DISCOUNTED CASH FLOW (D.C.F.)

This is a relatively new technique which recognises that *cost* values may be misleading unless related to *time*. Ignoring inflation, one pound money value today is effectively worth more than one pound payable at some time in the future. This is because the pound owned today can earn interest in the intervening period. Hence, the present day value of money to be received in the future should include the accrued amount of compound interest. This is expressed mathematically thus:

$$V_n = \frac{C_n}{(1 + r)^n}$$

where V = present value
C = cash flow
r = rate of discount (expressed as decimal fraction)
n = time period.

Example 1.8

a) Calculate the present value of £157 500 to be received in 5 years at a discount rate of 12% per annum.
b) Calculate the present value of £157 500 to be received annually in equal increments of £31 500 over 5 years at a 12% rate.
Solution:

a)
$$V_n = \frac{C_n}{(1 + r)^n}$$

$$\therefore V_5 = \frac{157\ 500}{(1.12)^5}$$
$$= \frac{157\ 500}{1.762}$$
$$= £89\ 387$$

b)
$$V_5 = \frac{C_1}{(1 + r)} + \frac{C_2}{(1 + r)^2} + \frac{C_3}{(1 + r)^3} + \frac{C_4}{(1 + r)^4} + \frac{C_5}{(1 + r)^5}$$
$$= \frac{31\ 500}{1.12} + \frac{31\ 500}{(1.12)^2} + \frac{31\ 500}{(1.12)^3} + \frac{31\ 500}{(1.12)^4} + \frac{31\ 500}{(1.12)^5}$$
$$= 28\ 125 + 25\ 112 + 22\ 422 + 20\ 019 + 17\ 874$$
$$= £113\ 552$$

In practice these values are not normally calculated and can be taken directly from discount tables published for unit values. As the name implies the D.C.F. method is based only upon cash considerations over a time period when cash flows in and when it flows out. One way of applying D.C.F. information is by using what is known as the Net Present Value (N.P.V.) method.

Net Present Value (N.P.V.)

Net present value can be defined as the present value of income arising from a project, less the present value of expenditure on the project, this being arrived at by discounting at a given rate of interest. An example will illustrate the method.

Example 1.9

A project will cost £12 000, and will return for each year over 6 years an income of £2 500, £3 500, £4 000, £6 500, £4 500 and £4 500 respectively. Interest charge (discount) rate on capital is 14%. Calculate the net present value (N.P.V.) using the discounted cash flow method.

Solution

Year	Cash Flow Out (£)	Cash Flow In (£)	Present Value Income (£)
0	12 000		
1		2 500	$\dfrac{2\ 500}{1.14} = 2\ 193$
2		3 500	$\dfrac{3\ 500}{(1.14)^2} = 2\ 693$
3		4 000	$\dfrac{4\ 000}{(1.14)^3} = 2\ 700$
4		6 500	$\dfrac{6\ 500}{(1.14)^4} = 3\ 849$
5		4 500	$\dfrac{4\ 500}{(1.14)^5} = 2\ 337$
6		4 500	$\dfrac{4\ 500}{(1.14)^6} = 2\ 050$
	$\overline{12\ 000}$ $\overline{25\ 500}$		$\overline{15\ 822}$

$$\therefore \text{ N.P.V.} = \text{present value of (income)} - \text{(expenditure)}$$
$$= £15\ 822 - £12\ 000$$
$$= £3\ 822$$

This example shows the N.P.V. value to be in surplus by £3 822. Hence, the cost of capital (i.e., 14%) to the company is easily covered by this project. If two projects, say were under consideration with only sufficient capital to undertake one, then comparison could be made using the N.P.V. figures as a basis for decision. Any project showing a N.P.V. loss would be considered unacceptable.

The figures shown in the 'Present Value – Income' column in the solution are calculated using the formulae given in the earlier explanation of the D.C.F. method, e.g., for year 4;

$$V_4 = \frac{6\ 500}{(1.14)^4} = £3\ 849$$

At the end of the last section (1.5) the question of machine replacement was raised. The use of the discounted cash flow method to determine the N.P.V. estimates of machines will help in determining an optimum replacement period.

Example 1.10

A machine tool is purchased for £40 000 using capital costing 16%. The estimated returns over 8 years are £5 000, £10 000, £10 000, £15 000, £20 000, £15 000, £10 000 and £7 000 respectively. Determine the optimum replacement period.

Solution

Year	Cash Flow Out (£)	Cash Flow In (£)	P.V. Income (£)	N.P.V. (£)
0	40 000			
1		5 000	$\dfrac{5\ 000}{1.16} = 4\ 310$	$-40\ 000 + 4\ 310 = -35\ 690$
2		10 000	$\dfrac{10\ 000}{(1.16)^2} = 7\ 432$	$-34\ 690 + 7\ 432 = -28\ 258$
3		10 000	$\dfrac{10\ 000}{(1.16)^3} = 6\ 407$	$-28\ 258 + 6\ 407 = -28\ 151$
4		15 000	$\dfrac{15\ 000}{(1.16)^4} = 8\ 284$	$-28\ 151 + 8\ 284 = -13\ 567$
5		20 000	$\dfrac{20\ 000}{(1.16)^5} = 9\ 522$	$-13\ 567 + 9\ 522 = -4\ 045$
6		15 000	$\dfrac{15\ 000}{(1.16)^6} = 6\ 157$	$-4\ 045 + 6\ 157 = +2\ 112$
7		10 000	$\dfrac{10\ 000}{(1.16)^7} = 3\ 538$	$2\ 112 + 3\ 538 = +5\ 650$
8		7 000	$\dfrac{7\ 000}{(1.16)^8} = 2\ 135$	$5\ 650 + 2\ 135 = +7\ 785$
		92 000	47 785	

The optimum replacement period will be when the N.P.V. value is zero, that being after year 5 when the N.P.V. changes from loss to surplus. This is when the D.C.F. income will have covered the return of the initial capital outlay.

1.7 ECONOMIC BATCH QUANTITIES

Much engineering production carried on in this country is of the type known as *batch production* (see Section 5.1). This occurs where the production of a component is not carried on continuously, but rather parts are produced in batches. The batches may be made to suit customers' requirements, as and when orders are placed, or may be made for stock. Most firms will choose the latter procedure for these two main reasons:

a) The stock acts as a buffer between input and output, and production batches can be arranged at periodic intervals to replenish the falling stock. These input time intervals will be independent of the customers' withdrawals from stock. This will make scheduling problems on the shop floor somewhat easier.

b) The stock ensures that customers do not have to wait an unacceptable time before delivery of the order.

Once the method of manufacture has been decided, and all the costs have been evaluated including cost of carrying the stock, then it becomes possible to choose which batch quantity should be produced to give minimum total cost/piece. It should be noticed here that break-even charts are best constructed to show total costs at any quantity Q to be manufactured. It will be seen however that the more involved cost analysis required for economic batch quantities is best done upon a chart showing total costs/piece at any quantity Q to be manufactured. This is illustrated in Fig 1.13(a) and (b), where Fig 1.13(a) shows fixed and variable costs for a process as total costs (as in break-even charts), and Fig 1.13(b) shows the same costs for the same process as total costs/piece.

The variable costs which rise uniformly in Fig 1.13(a), are a constant value per piece for any quantity Q, and are denoted by means of a horizontal straight line in Fig 1.13(b). Hence the total variable cost $V = aQ$. The fixed costs/piece in Fig 1.13(b) are denoted by means of a falling line and at any quantity Q are equal to $\dfrac{F}{Q}$.

In the case of batch quantities, the cost/piece of storage should also be included, and will include the interest cost of the capital invested in the articles kept in stock, stores overheads, stores personnel costs and the cost of deterioration of the articles. These are known as storage carrying costs. The essence of the theory of economic batch quantities

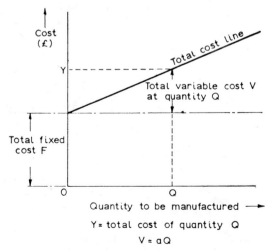

Fig 1.13(a)   Total Cost Chart.

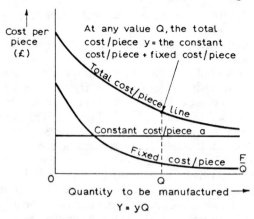

At any value Q, the total
cost/piece y = the constant
cost/piece + fixed cost/piece

Fig 1.13(b)   Total Cost/piece Chart.

is that storage carrying costs/piece increase as quantity Q increases, and
hence the total cost/piece line will begin to rise again beyond some point
where total costs/piece are a minimum. This theory is not acceptable to
everyone, but much interesting literature has been written on the subject.

If the storage carrying costs behave as described above then the total
cost picture will appear as shown in Fig 1.14. The carrying costs/piece

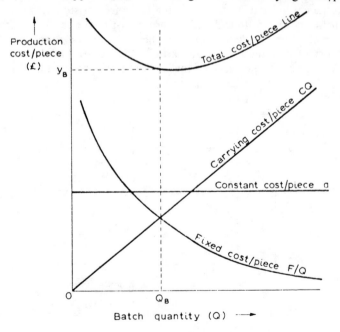

Fig 1.14   Total Cost/piece Chart.

are denoted by a uniformly rising line in Fig 1.14 and are equal to CQ where C is the carrying cost factor. It is assumed then that the carrying costs increase proportionally as Q increases, and hence the total cost/piece is minimum at some quantity Q_B. Quantity Q_B is called the *economic batch quantity*. This is the batch size which should be chosen for production before turning the machine over to the production of some other product. It will be found easier, once the cost information is known, to calculate Q_B, rather than attempt to draw the chart to scale.

$$\text{Total cost/piece} = \text{constant cost/piece} + \text{fixed cost/piece} + \text{carrying cost/piece}$$

$$\therefore y = a + \frac{F}{Q} + CQ$$

At point Q_B, y must be a minimum, therefore by calculus

$$\frac{dy}{dQ} = 0$$

$$\text{where } \frac{dy}{dQ} = \frac{-F}{Q^2} + C$$

$$\therefore \frac{-F}{Q^2} + C = 0 \quad \therefore Q^2 = \frac{F}{C}$$

$$\therefore Q = \sqrt{\frac{F}{C}} = Q_B \text{ at the minimum}$$

It will be noted that the value of the constant costs/piece a, does not affect the value.

The evaluation of the storage carrying cost factor C is complex and a text book on economic batch quantity theory should be consulted for detailed information.

Example 1.11
Find the economic batch quantity Q_B for (a) the automatic and (b) the capstan lathe, given the cost information for each machine shown in Example 1.3. It is known that the cost carrying factor C is equal in both cases to £0·0025/piece. What is the total cost/piece of production in each case?

Solution

(a) $F = 292{\cdot}00$, $C = 0{\cdot}0025$, and $a = 0{\cdot}44$

$$\therefore Q_B = \sqrt{\frac{F}{C}} = \sqrt{\frac{292{\cdot}00}{0{\cdot}0025}} = \sqrt{116\,800} = 342$$

Total cost/piece of producing 342 components on the automatic

$$= y_B = a + \frac{F}{Q_B} + CQ_B$$

$$= 0.44 + \frac{292}{342} + (0.0025 \times 342)$$

$$= 0.44 + 0.85 + 0.85$$

$$= £2.14 \text{ per piece (Note that } \frac{F}{Q} = CQ \text{ at quantity } Q_B.)$$

b) $F = 41.50$, $C = 0.0025$ and $a = 1.09$

$$\therefore Q_B = \sqrt{\frac{F}{C}} = \sqrt{\frac{41.50}{0.0025}} = \sqrt{16\ 600} = 129$$

Total cost/piece of producing 129 components on the capstan lathe

$$= y_B = a + \frac{2F}{Q_B} = 1.09 + \left(\frac{2 \times 41.50}{129}\right)$$

$$= 1.09 + 0.64$$
$$= £1.73 \text{ per piece}$$

The reader must not get confused between the parameters, break-even quantity Q_E, and the minimum economic batch quantity Q_B. The former is the batch quantity where the total costs (and also the total cost/piece) for two or more equally feasible processes are equal. The latter is the batch quantity where the total cost/piece for *one particular process* is at a minimum. The relationship between Q_E and Q_B in Examples 1.3 and 1.11 is shown in Fig 1.15.

The break-even quantity Q_E will still be equal to 385, and will not have been modified by the inclusion of the carrying costs, as shown below.

The total cost/piece at the break-even point $Q_E = y_E$

For the automatic $\quad y_E = 0.44 + \dfrac{292.00}{Q_E} + (0.0025 \times Q_E)$

For the capstan lathe $y_E = 1.09 + \dfrac{41.50}{Q_E} + (0.0025 \times Q_E)$

$$\therefore \frac{292.00}{Q_E} - \frac{41.50}{Q_E} = 1.09 - 0.44$$

$$\therefore \frac{250.50}{Q_E} = 0.65$$

$$Q_E = \frac{250.50}{0.65} = 385$$

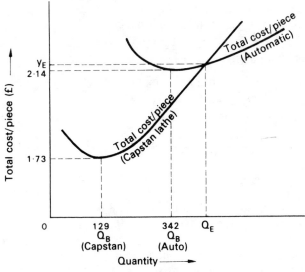

Fig 1.15   Cost Chart.

The total cost/piece at quantity $Q_E = y_E$

$$\text{where } y_E = 1{\cdot}09 + \frac{41{\cdot}50}{385} + (0{\cdot}0025 \times 385)$$

$$= 1{\cdot}09 + 0{\cdot}11 + 0{\cdot}96$$
$$= £2{\cdot}16 \text{ per piece}$$

Example 1.12
The production of an article using a multi-die press expended fixed costs equal to £9 000. The constant costs for each article were:
Material cost £1·00 for 30
Labour cost   £1·00 for 120
Overheads   £1·00 for 40.
The carrying costs factor was estimated as £1·00 for 200 000.
a) Calculate the economic batch quantity.
b) Calculate the total cost of the production run.
c) After production commenced, it was found that the carrying costs factor would need to be increased by 10% to be reasonably accurate. Calculate the revised total cost.

Solution

a) $F = 9\,000$

          $C = 0{\cdot}000\,005$

$$\therefore Q_B = \sqrt{\frac{F}{C}} = \sqrt{\frac{9\,000}{5 \times 10^{-6}}} = \sqrt{18 \times 10^8} = 4{\cdot}243 \times 10^4 = 42\,430$$

b) $a = \frac{1}{30} + \frac{1}{120} + \frac{1}{40} = \frac{8}{120} = \frac{1}{15}$

$$\therefore \, y_B = a + \frac{2F}{Q_B} = \frac{1}{15} + \left(\frac{2 \times 9\,000}{42\,430}\right)$$

$$= 0{\cdot}066\,7 + 0{\cdot}424\,3 = £0{\cdot}491 \text{ per piece}$$

$$Y_B = \text{total cost of producing quantity } Q_B = y_B Q_B$$

$$= 0{\cdot}491 \times 42\,430 = £20\,800$$

c) New $C = 0{\cdot}000\,005\,5$

$$\text{New } Q_B = \sqrt{\frac{9\,000}{5{\cdot}5 \times 10^{-6}}} = 40\,440$$

$$\text{New } y_B = 0{\cdot}066\,7 + \left(\frac{2 \times 9\,000}{40\,440}\right)$$

$$= 0{\cdot}066\,7 + 0{\cdot}445\,1 = £0{\cdot}511\,8 \text{ per piece}$$

New $Y_B = 0{\cdot}511\,8 \times 40\,440 = £20\,700.$

From this example it can be seen that an increase in the carrying costs will cause a decrease in the economic batch size, but the total cost of production of Q_B remains approximately the same.

It must not be thought that values of Q_E are rigid once calculated. Even assuming that the cost values obtained from the accountant are reasonably accurate, the production engineer might well be forced to ignore the information that break-even charts give him.

For example, an order for 600 components might be placed with the manufacturing company who produce the articles costed in Example 1.11. The production manager might wish to use an automatic to produce the order as the break-even quantity Q_E is equal to 385, and the cost of producing on the capstan lathe rises sharply above the automatic production cost, beyond the value of Q_E. But if the automatics are tied up on other more important work on a long production run, then the order of 600 might have to be placed on a capstan lathe to meet the delivery date. Goodwill gained here might lead to larger orders later. Say a reasonable price of £3·00/piece was charged, then the profit situation can be depicted on a break-even chart as shown in Fig 1.16.

The sales line can be drawn first by computing the price of 600 articles, which is £1 800. The sales line then is a straight line drawn through the ordinates (0,0) and (600,1 800).

From this chart it can be seen that total profit Z_C from the batch of 600 produced upon the capstan lathe is equal to £204. Total profit Z_A from the batch of 600 produced on the automatic is equal to £342. The loss in profit is £138. This might be a small price to pay when all factors are taken into consideration.

In the case of a calculated value of Q_B, it can be seen from the examples given that the total cost curve does not rise very sharply after the value of Q_B. Therefore deviations from this value are not of vital importance as the following example shows.

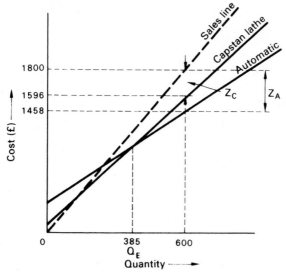

Fig. 1.16 Cost—Profit Chart.

Example 1.13

In example 1.12, as a result of a 10% increase in the carrying cost factor, the value of Q_B changed from 42 430 to 40 440. What would the percentage increase in total cost be if the original value of Q_B were to be produced?

Solution

Total cost Y_B of producing 40 440 components = £20 700
Total cost Y_B of producing 42 430 components = £0·511 8 × 42 430
                                                                        = £21 716

$$\% \text{ increase in total cost} = \frac{21\ 716 - 20\ 700}{20\ 700} \times 100 = 4·9\%$$

i.e. an increase of 10% in C only increases Y by 4·9%.

If scheduling difficulties on the shop floor make it inconvenient to produce batches of exactly 40 440 by the chosen process then obviously 40 440 must not be allowed to become a sacred value from which it is impossible to deviate, but should be changed to suit practical circumstances. It would seem preferable to increase Q_B rather than decrease it. If the carrying costs do not increase in the manner suggested earlier, then the production engineer's practical instincts might tell him that once he has a batch to produce which is larger than Q_B, he should carry on producing as long as possible before breaking the job down. In effect he might visualize the cost situation as Fig 1.17, with no turning point on the total cost curve, i.e., the more that can be produced at one setting the better provided anticipated future requirements will allow this.

In any event it is suggested that a highly accurate value is not needed for Q_E or Q_B where sophisticated and expensive costing techniques are required in order to obtain the necessary information, but rather that a cruder value will serve just as well to give the production department a 'sighting shot' at the target. Some authorities on the subject might go further and assert that to obtain an accurate value for the carrying cost factor is impossible, or at best to obtain an approximate assessment of the value would involve great expense. Therefore it is better to ignore it, and not to base the manufacturing programme on the principle of economic batch quantities produced at a particular frequency.

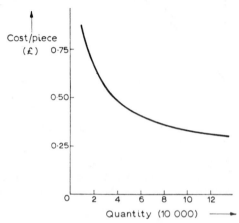

Fig 1.17   Cost Chart.

Group Technology

In the field of production engineering, a recent innovation has been the introduction of a technique called group technology. In some instances it has been found that, when this technique is used in organising work flow through a manufacturing shop, then work batch sizes deduced by the theory of 'Economic Batch Quantities' are not sensible.

Group technology is, in effect, the replacing of the traditional manufacturing shop floor organisation by the grouping of similar types of work into families. In addition, machine tools are organised into groups to manufacture these families of work on a flow-line principle (see Chapter 5). A variety reduction programme (see Section 5.1) is the first step towards group technology.

1.8 ECONOMICS OF METAL REMOVAL

When a manufacturing process consists of removing metal with a single point tool, the type of tool used or cutting speed chosen can have an effect upon the total cost of the product. It is worth considering this because the removal of metal in this manner is still a major process in the engineering industry. In a roughing operation the object is to remove a certain volume of material at minimum cost or minimum time, or maximum profit, and the type of tool and cutting speed should be chosen accordingly. In a finishing operation the object is to improve a certain area of material until it is of the desired quality of finish. In the following discussion the chosen criterion is the removal of certain volume of material at mimimum cost. Again it should be emphasized that the analysis used to obtain the optimum conditions is worthless unless the cost information used is relatively accurate.

F. W. Taylor introduced the well known relationship between the cutting speed used in a metal removing operation. and the life of the tool, viz.,

$$VT^n = C$$

where $V =$ cutting speed
$T =$ tool life

(Although in basic SI units metres and seconds should be used, metres and minutes are the practical units)

$n =$ an index closely related to the cutting tool material, and the following values may be used:
0·1 to 0·15 for high speed steel tools
0·2 to 0·4 for tungsten carbide tools
0·4 to 0·6 for ceramic tools
$C =$ a constant

This is an empirical relationship and for any given set of cutting conditions over a practical range of speeds, it can be considered valid. If the cutting conditions are changed however, (i.e., feed, depth of cut, rake angle, tool shape, workpiece material, etc.) then the relationship will cease to be true. It can be seen then that for a particular machining operation all the variables, other than V and T, must be kept constant otherwise the law is not valid. It can be shown on a graph as illustrated in Fig 1.18.

The curve is exponential. Cutting tests must be used to obtain values of n and C. These values are difficult to obtain accurately, because in

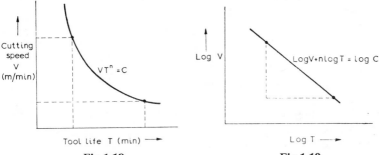

Fig 1.18   Fig 1.19

turn it is difficult to assess when the effective cutting life of the tool has ceased during the test. As a tool is tested with varying cutting speeds, a sensible criterion must be adopted to determine tool life. Then the values obtained for V and T from the test with controlled cutting conditions can be plotted on a graph, Fig 1.19, using a log-log scale.

The slope of the straight line will give the value of n, and hence a value for C can be obtained.

It can be seen that if cutting speed V is increased, then tool life T will decrease. Hence, metal is removed faster and therefore more cheaply. But tool life is shorter and therefore tools replacement and servicing are more costly. This cost situation is shown in Fig 1.20.

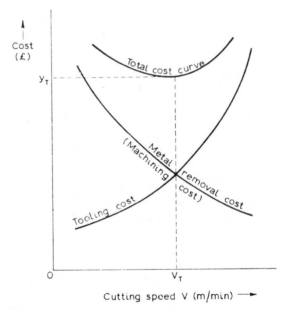

Fig 1.20 Cost Chart.

V_T = Optimum cutting speed where the total cost of machining a batch of components y is at a minimum.

In order to find an expression for V_T the tooling cost and metal removal cost (or machining cost) must be added to give the total cost. Then by calculus the turning point of the curve and hence V_E can be found.

Let H = machining cost/minute i.e., labour cost/minute + overheads/ minute.

Let J = tooling cost i.e., cost of changing tool + cost of regrinding + tool depreciation.

Let y_1 = cost of machining metal/unit volume of metal cut.

Let y_2 = cost of servicing tools/unit volume of metal cut.

Let y = total cost/unit volume of metal cut = $y_1 + y_2$

Time to machine a unit volume of metal in minutes

$$= \frac{1}{dfV}$$

where d = depth of cut

f = feed in length/rev

V = cutting speed

But as stated earlier, d and f are constants therefore $\dfrac{1}{dfV} = \dfrac{K}{V}$ where K is a constant.

$$\therefore y_1 = \frac{HK}{V}$$

The number of tool changes in $\dfrac{K}{V}$ minutes = $\dfrac{K}{TV}$ where T is tool life in minutes at cutting speed V.

$$\therefore y_2 = \frac{JK}{TV} \qquad \text{But } T = \left(\frac{C}{V}\right)^{\frac{1}{n}}$$

$$\therefore y_2 = \frac{JK}{\left(\dfrac{C}{V}\right)^{\frac{1}{n}} \times V} = \frac{JK(V)^{\frac{1-n}{n}}}{C^{1/n}}$$

$$\therefore y = y_1 + y_2 = \frac{HK}{V} + \frac{JK(V)^{\frac{1-n}{n}}}{C^{1/n}}$$

Differentiating

$$\frac{dy}{dV} = \frac{-HK}{V^2} + \left(\frac{1-n}{n}\right)\frac{JK}{C^{1/n}}\left(V\right)^{\frac{1-2n}{n}}$$

For a minimum $\dfrac{dy}{dV} = 0$

$$\therefore \frac{H}{V^2} = \left(\frac{1-n}{n}\right)\frac{J}{C^{1/n}}\left(V\right)^{\frac{1-2n}{n}} \quad (K \text{ cancelling out})$$

$$\therefore H = \left(\frac{1-n}{n}\right)\frac{J}{C^{1/n}}\left(V\right)^{\frac{1}{n}} = \left(\frac{1-n}{n}\right)J\left(\frac{V}{C}\right)^{\frac{1}{n}}$$

Now if R = the ratio $\dfrac{\text{tooling costs}}{\text{machining costs}}$ then $R = \dfrac{J}{H}$

$$\therefore \frac{1}{R} = \left(\frac{1-n}{n}\right)\left(\frac{V}{C}\right)^{\frac{1}{n}}$$

$$\therefore \left(\frac{V}{C}\right)^{\frac{1}{n}} = \frac{n}{R(1-n)} \qquad \therefore V = C\left[\frac{n}{R(1-n)}\right]^{n}$$

$$= V_T \text{ at the minimum}$$

This expression will enable V_T to be calculated so that the optimum cutting speed can be found to give minimum cost y_T for the batch. It should be noticed that n from Taylor's equation is important in this equation, hence the need to obtain its value accurately as described earlier. In this analysis we have not included the costs of handling the tool.

Example 1.14(a)
The following information is known for a lathe machining operation on a particular component:

$$n = 0\cdot25$$
$$c = 150$$

Tool change time	= 6 minutes
Tool regrind time	= 5 minutes
Machine running cost	= £0·025/minute
Depreciation of tool/regrind	= £0·125

Calculate the optimum cutting speed V_T

Solution

Tool change cost	= 6 × £0·025 = £0·15
Tool regrind cost	= 5 × £0·025 = £0·125
Tool depreciation	= £0·125

$$\therefore R = \frac{\text{tooling costs}}{\text{machining costs}} = \frac{0\cdot15 + 0\cdot125 + 0\cdot125}{0\cdot025} = 16$$

$$V_T = 150 \left(\frac{4}{4 \times 16 \times 3}\right)^{\frac{1}{4}} = 150 \left(\frac{1}{48}\right)^{\frac{1}{4}} = 57 \text{ metres per minute}$$

Example 1.14(b)
Using the information of problem 1.14(a) previously, if 200 components are required with a machined length of 400 mm at 100 mm diameter and a feed of 0·25 mm/rev, calculate the time required to machine 200 parts, and the total cost y_T of machining them (ignoring handling costs of tools).

Solution

$$\text{Spindle speed} = \frac{V}{\pi d} = \frac{57\,000}{3\cdot142 \times 100} = 181 \text{ rev/min}$$

$$\text{Machining time/piece} = \frac{\text{length to be machined}}{\text{length fed in one minute}}$$

$$= \frac{400}{181 \times 0\cdot25} = 8\cdot84 \text{ min}$$

$$\text{Tool life } T \text{ at 57 m/min cutting speed} = \left(\frac{C}{V}\right)^{\frac{1}{n}} = \left(\frac{150}{57}\right)^{4}$$

$$= 48 \text{ minutes}$$

$\therefore$ Number of parts/regrind $= \dfrac{48}{8\cdot84} \qquad = 5\cdot43$

Tool change cost/piece $= \dfrac{0\cdot15}{5\cdot43} \qquad = £0\cdot027\ 6$

Tool regrind cost/piece $= \dfrac{0\cdot125}{5\cdot43} \qquad = £0\cdot023\ 0$

Tool depreciation/piece $= \dfrac{0\cdot125}{5\cdot43} \qquad = £0\cdot023\ 0$

Machining cost/piece $= 8\cdot84 \times £0\cdot025 = £0\cdot221\ 0$
$$\overline{}$$

$\therefore$ Total cost/piece y_T (by addition) $\qquad = £0\cdot294\ 6$

Total cost of batch of 200 $= Y_T \qquad = 200 \times £0\cdot294\ 6 = £58\cdot92$

Production rate/hour $=$

$$\frac{60}{\text{machining time/piece} + \text{change time/piece} + \text{regrind time/piece}}$$

$$= \frac{60}{8\cdot84 + \dfrac{6}{5\cdot43} + \dfrac{5}{5\cdot43}} = \frac{60}{10\cdot86} = 5\cdot52$$

Time required/200 parts $= \dfrac{200}{5\cdot52} = 36\cdot2$ hours

If the cutting speed is raised above V_T then the actual machining time/piece is less, but the total cost for the batch will then be greater than Y_T. If the cutting speed is reduced below V_T, then the actual machining time/piece is greater, and again the total cost will be greater than Y_T.

Consider Example 1.15, which illustrates this point.

Example 1.15

The information for a lathe machine operation is exactly as in Example 1.14, but the optimum cutting speed V_T is raised by 25%. What will the time now be to machine 200 parts and what will be the percentage increase in total cost/piece y.

Solution

$$\text{New } V = 57 + \frac{57}{4} = 71\cdot25 \text{ m/min}$$

$$\text{Spindle speed} = \frac{71\ 250}{3\cdot142 \times 100} = 227 \text{ rev/min}$$

$$\text{Machining time/piece} = \frac{400}{227 \times 0\cdot25} = 7\cdot05 \text{ min}$$

$$\text{New } T = \left(\frac{150}{71\cdot25}\right)^4 = 19\cdot6 \text{ min}$$

$$\text{No. of parts/regrind} \quad = \frac{19 \cdot 6}{7 \cdot 05} = 2 \cdot 78$$

$$\text{Tool change cost/piece} \quad = \frac{0 \cdot 15}{2 \cdot 78} = £0 \cdot 054\ 0$$

$$\text{Tool regrind cost/piece} \quad = \frac{0 \cdot 125}{2 \cdot 78} = £0 \cdot 045\ 0$$

$$\text{Tool depreciation/piece} \quad = \frac{0 \cdot 125}{2 \cdot 78} = £0 \cdot 045\ 0$$

$$\text{Machining cost/piece} = 7 \cdot 05 \times 0 \cdot 025 = £0 \cdot 176\ 3$$

$$\text{Total cost/piece } y \text{ (by addition)} \quad = £0 \cdot 320\ 3$$

$$\text{Percentage increase} = \frac{0 \cdot 320\ 3 - 0 \cdot 294\ 6}{0 \cdot 294\ 6} \times 100 = 8 \cdot 7\%$$

$$\text{Total cost of batch } Y = 200 \times 0 \cdot 320\ 3 = £64 \cdot 06$$

$$\text{Production rate/hour} = \frac{60}{7 \cdot 05 + \dfrac{6}{2 \cdot 78} + \dfrac{5}{2 \cdot 78}} = \frac{60}{11 \cdot 01} = 5 \cdot 46$$

$$\text{Time required/200 parts} = \frac{200}{5 \cdot 46} = 36 \cdot 6 \text{ hours}$$

Comparing Examples 1.14 and 1.15 it can be seen that a 25% increase in cutting speed increases the cost of the batch by 8·7%. Again, the theoretical optimum values should not be rigidly adhered to if practical circumstances dictate that an alternative value would be better. In the example quoted $n = \frac{1}{4}$, which indicates the use of a cemented carbide tool, therefore a rise in V from 57 m/min to 71·25 m/min is feasible.

It was stated at the beginning of Section 1.8 that the type of tool used OR the cutting speed chosen could have an effect upon the total cost. This has been demonstrated with respect to the cutting speed; in Example 1.16 the effect upon total cost of a change in the type of tool will be considered.

Example 1.16

In Examples 1.14 and 1.15 the tool used was a tungsten carbide tipped tool with the tip brazed to the tool shank. Calculate the total cost Y_T of machining 200 parts if a 'throwaway' tipped tool is used having eight available cutting edges. Also calculate the time required in this case to produce 200 parts.

Assume the following changes:

Tool change time now 2 minutes average. This allows for changing to fresh cutting edges which will take less than 2 minutes, and changing to a new tip when the tip is worn out, which will take more than 2 minutes. Tool regrinding time and hence tool regrinding cost now nil.

$$\text{Depreciation/tip edge} = £0 \cdot 125$$

Solution

$$\text{Tool change cost } = 2 \times £0 \cdot 025 = £0 \cdot 05$$
$$\text{Tool regrind cost } = 0$$
$$\text{Tool depreciation } = £0 \cdot 125$$

$$\therefore R = \frac{\text{tooling cost}}{\text{machining cost}} = \frac{0 \cdot 05 + 0 + 0 \cdot 125}{0 \cdot 025} = \frac{0 \cdot 175}{0 \cdot 025} = 7$$

$$V_T = 150\left[\frac{4}{4 \times 7 \times 3}\right]^{\frac{1}{4}} = 150\left(\frac{1}{21}\right)^{\frac{1}{4}} = 70 \cdot 0 \text{ m/min}$$

Spindle speed $=$	$\dfrac{70\,000}{3 \cdot 142 \times 100}$	$= 223$ rev/min
Machining time/piece $=$	$\dfrac{400}{223 \times 0 \cdot 25}$	$= 7 \cdot 16$ min
Tool life T $=$	$\left(\dfrac{150}{70}\right)^4$	$= 21 \cdot 1$ min
Number of parts/tip edge $=$	$\dfrac{21 \cdot 1}{7 \cdot 16}$	$= 2 \cdot 95$
Tool change cost/piece $=$	$\dfrac{0 \cdot 05}{2 \cdot 95}$	$= £0 \cdot 0169$
Depreciation/piece $=$	$\dfrac{0 \cdot 125}{2 \cdot 95}$	$= £0 \cdot 0424$

Machining cost/piece $= 7 \cdot 16 \times £0 \cdot 025 = £0 \cdot 1790$
Total cost/piece y_T (by addition) $= \overline{£0 \cdot 2383}$
Total cost of batch $Y_{T'}$ $= 200 \times £0 \cdot 2383 = £47 \cdot 66$

$$\text{Production rate/hour} = \frac{60}{7 \cdot 16 + \dfrac{2}{2 \cdot 95}} = \frac{60}{7 \cdot 84} = 7 \cdot 65$$

$$\text{Time required/200 parts} = \frac{200}{7 \cdot 65} = 26 \cdot 1 \text{ hours}$$

When these values are compared with those obtained in Example 1.14(b) it is seen that, using throw-away tips, the total cost Y_T is considerably less and 10·1 hours are saved on the total time required to complete the batch. This explains in some part the increasing popularity of this type of tool arrangement.

1.9 ECONOMICS OF MATERIAL UTILIZATION

Another field of manufacturing which will yield results from an examination of the costs involved is the choice and utilization of material. The modern technique of *value analysis* is largely concerned with this field because a change in the choice of materials, and/or a change in the method of utilizing the material, can make spectacular cost savings. Several materials may offer the required properties in a particular component. The engineering choice has never been wider for designers particularly if one includes the whole range of plastics.

In this section the economics of material utilization will be examined. *Material utilization* can be defined as the ratio:

$$m = \frac{\text{weight of the component}}{\text{weight of material used to make component}}$$

The nearer the value of this ratio approaches unity, the less waste will be incurred and hence the smaller will be the variable cost for the manufacturing process. Figures 1.21, 1.22 and 1.23 show examples of how the value of the ratio *m* has been increased by a change in method.

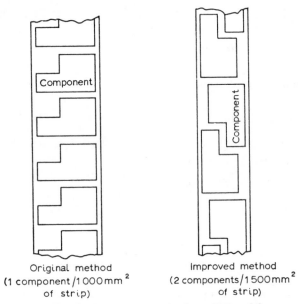

Original method
(1 component/1 000 mm^2
of strip)

Improved method
(2 components/1 500 mm^2
of strip)

Fig 1.21   Improved Utilization of Material.

Figure 1.21. The component is produced from sheet steel by a blanking operation on a press. The amount of scrap strip left can be reduced by a better arrangement of blanks punched from the strip.

Figure 1.22. The component is machined from a casting. The amount of scrap can be reduced by coring a hole in the casting.

Figure 1.23. The component is machined from 40 mm diameter bright bar in the original method. In the improved method it is fabricated by brazing a 40 mm diameter collar on to a 16mm diameter rod.

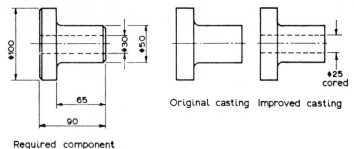

Original casting   Improved casting

Required component

Fig 1.22 Improved Utilization of Material.

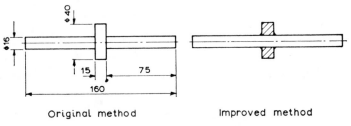

Original method   Improved method

Fig 1.23 Improved Utilization of Material.

In each case illustrated the change in method involves a more costly primary process. In Fig 1.21 for example a dearer blanking tool and probably a larger press would be required for the improved method. Hence a break-even analysis would be required as Fig 1.24, and the final choice of method would depend upon the quantity required.

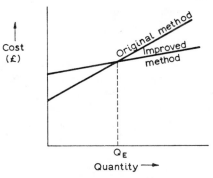

Fig 1.24 Break-even Chart.

The ratio m gives a quick, general idea of material utilization, but where many components are under consideration it does not allow a very useful comparison to be made. A value of m equal to 0·25 for component *A* might be much less serious than a value of *m* equal to 0·80 for component *B*. This would be the case if *A* was made from cheap material and *B* was made from a very expensive material where any waste would be very costly. Hence, cost again becomes the criterion. It has been suggested that an *equivalent material utilization* analysis should be carried out for a series of components (which might or might not form an assembly) where the ratio *m* is weighed against the total cost of production *Y*. This seems to be the sensible approach, because all the components are in effect reduced to a common denominator. The information is best tabulated as shown below where it can be assumed that four components make up an assembly.

Col. 1	Col. 2	Col. 3	Col. 4	Col. 5
		Cost of	Relative	Partial
Component No.	Material Utilization	Production/ piece	Production Cost	Utilization value
1	m_1	y_1	$X_1 = \dfrac{y_1}{\Sigma y}$	$m_1 X_1$
2	m_2	y_2	$X_2 = \dfrac{y_2}{\Sigma y}$	$m_2 X_2$
3	m_3	y_3	$X_3 = \dfrac{y_3}{\Sigma y}$	$m_3 X_3$
4	m_4	y_4	$X_4 = \dfrac{y_4}{\Sigma y}$	$m_4 X_4$
	Total Production Cost of Assembly = Σy		Total = 1	$M = \Sigma mx$

The total of column 5 is ΣmX which is called the *equivalent material utilization value* M for the assembly. If each individual value *m* is improved to a value of 1 then M will equal 1. If the value of X in column 4 is compared to the value of mX in column 5 for each component, it becomes much clearer which components should first be studied in order to make a worthwhile saving. This can best be shown by means of an example.

Example 1.17
The material utilization ratios for five components comprising an assembly are 0·2, 0·8, 0·5, 0·2 and 0·35, and the total costs of production are £3·00, £2·10, £1·10, £0·60 and £1·00 respectively. Calculate the equivalent material utilization value M, and state which component(s) could most profitably be studied with a view to improving M.

Solution

Compt. No.	m	$(£)y$	X	mX
1	0·2	3·00	0·384	0·077
2	0·8	2·10	0·269	0·215
3	0·5	1·10	0·141	0·071
4	0·2	0·60	0·078	0·016
5	0·35	1·00	0·128	0·045
		7·80	1·000	0·424

$$M = 0·424$$

For each component, X represents a value which could be attained if $m = 1$ (i.e., perfection from the material utilization point of view). mX represents a value which can be improved (by improving m) to a theoretically possible maximum equal to X.

The component which would most repay study is component No. 1 as $7 \cdot 7\%$ could be improved to a possible maximum of $38 \cdot 4\%$.

This information can be shown pictorially and more clearly on a graph as illustrated in Fig 1.25.

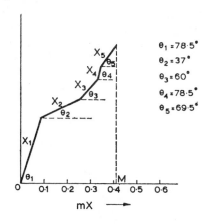

Fig 1.25

Scale each value of X at an angle θ to the horizontal axis where $m = \text{Cos } \theta$. Therefore X_1 is drawn to scale $= 0 \cdot 384$ at an angle $\theta_1 = 78 \cdot 5°$ approx. Complete the graph by drawing each successively at angle θ. Then it can be seen that the component with the longest line and largest angle will best repay improving its material utilization. This can be seen to be component No. 1.

The above method assumes that the proportion of material costs to production costs is approximately the same for each component. Where this is not so it would be better to use the ratio (cost of material/piece) to the (total cost/piece) for column 3. Also if this analysis is applied to every component made on the shop floor (as opposed to an assembly of components), then the batch quantity Q must also be taken into account. If Q is small then it will not be worth altering the method in order to improve m. Consider Example 1.18.

Example 1.18
Assume that the five components in Example 1.17 are individual components made of a similarly priced steel, and produced in the following quantities/week: 100, 1 000, 5 000, 1 000, and 10 000. Which components(s) now would most repay study for improving the value of m?

Solution

Compt. No.	Q	m	y	Qy	$X = \dfrac{Qy}{\Sigma Qy}$	mX
1	100	0·2	3·0	300	0·016	0·003
2	1 000	0·8	2·1	2 100	0·114	0·091
3	5 000	0·5	1·1	5 500	0·297	0·149
4	1 000	0·2	0·6	600	0·033	0·006
5	10 000	0·35	1·0	10 000	0·540	0·189
				18 500	1·000	M = 0·438

The new value of $M = 0\cdot438$

The component now which would most repay study is component No. 5 as 18·9% could be improved to a possible maximum of 54%. This is very clearly shown on the graph Fig 1.26 remembering the rule: 'largest angle and side most repay study'. The values of θ remain the same as Fig 1.25.

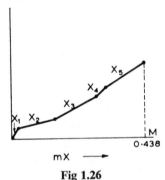

Fig 1.26

Again it might be stated that if the method of producing component No. 5 was altered in order to improve m, then total cost/piece y would also be altered. This would necessitate some consideration as mentioned earlier in this section.

1.7 ESTIMATING STANDARD TIMES

In order for accurate labour costs to be estimated prior to production, it is first necessary to accurately estimate *standard times* for carrying out a certain operation or process. From BS 3138: 1969, A Glossary of Terms in Work Study, *standard time* is defined as the total time in which a job should be completed at standard performance. It includes an allowance for relaxation known as R.A. (relaxation allowance), which in turn includes a fatigue allowance. From the same BS Specification, *standard performance* is defined as the rate of output which qualified workers will naturally achieve without over-exertion as an average over the working day or shift provided they know and adhere to the specified method and provided they are motivated to apply themselves to the work. In order to evaluate the standard time for a job so that cost analysis may be carried out before the job is put into production, *Work Study* techniques are required. In this section only a brief outline may be given.

Work carried out on a particular workpiece will consist of all manipulative human operations, or part manipulative operation and part machine operation, or all machine operation. The methods used to estimate the times of such operations are given below.

Estimating manipulative operation times

This is done using synthetic data which is defined as tables and formulae derived from the analysis of accumulated work measurement data, arranged in a form suitable for building up standard times, machine process times, etc. by synthesis. Many firms collect their own synthetic data for a particular process with which they are primarily concerned from past records of time studies. The time studies are taken with a stop watch after production has started in order to allow a wage rate to be fixed for the job. In order to estimate a standard time by synthesis, the job must be broken down into separate elements. The time for each element is obtained from the tables, the separate element times then totalled, and finally allowances added. The final result will not be as accurate as a stop watch time but will be quite adequate.

Many systems of synthetic data both for special or general work have been published. These are called *Predetermined Motion Time Systems*. The two best known are probably the *Work Factor* system and the *Methods Time Measurement* system. As with all synthesis, skill is needed to obtain accurate results, and training is needed in order to apply predetermined motion time systems.

Estimating machining operation times

With this type of operation where a machine is required in order to complete the work, the machine will most usually be human operated. In this case the standard time will consist of part manipulative time and part machining time. In the case of a fully automatic machine the standard time will be all machining time.

Formulae and data for calculating machining times can be found in any text books on Machine Tools, but a few are given here in order to allow the principle to be grasped. They are grouped under two broad headings, (a) rotary machining operations, and (b) reciprocating machining operations.

a) *Rotary machining operations* (such as turning, drilling, milling, etc.)

Spindle speed S (rev/min) $= \dfrac{1\,000V}{\pi d}$ where $V =$ cutting speed of cutter or workpiece in metres per minute and $d =$ work or cutter diameter in millimetres

Time for one cut T (minutes) $= \dfrac{L}{Sf}$ where L = length of cut (mm) and f = feed (mm/rev).

In the case of milling an allowance called the approach distance A must be added to the length of the machined surface L such that $L = A + l$. This is illustrated in Fig 1.27(a) and 1.27(b) for peripheral and face milling.

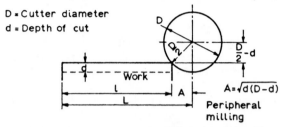

Fig 1.27(a)   Approach Distance for Peripheral Milling.

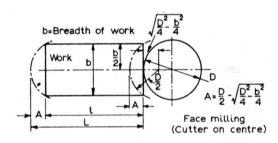

Fig 1.27(b)   Approach Distance for Face Milling.

In both these cases $T = \dfrac{L}{f} = \dfrac{l + A}{f}$ where f = feed (mm/min).

b) *Reciprocating machining operations* (such as shaping, planing, etc.) With this type of machine one cycle of work consists of a forward stroke, (of tool or work), and a return stroke usually at a faster speed to reduce idle time.

$\therefore$ Time to complete one cycle $= \dfrac{L}{V_1} + \dfrac{L}{V_2} = L\left(\dfrac{1}{V_1} + \dfrac{1}{V_2}\right)$

where V_1 = average forward cutting speed (m/min)

and V_2 = average return cutting speed (m/min)

Hence, the time to take one cut over work of breadth b, length L using a feed of f (mm/stroke) is:

$$T = \frac{\text{width of work}}{\text{feed/stroke}} \times \text{time to complete one cycle}$$

$$= \frac{b}{f} \times L\left(\frac{1}{V_1} + \frac{1}{V_2}\right) = \frac{bL}{f}\left(\frac{1}{V_1} + \frac{1}{V_2}\right)$$

One example will suffice to illustrate the method of obtaining a standard time (part manipulative time and part machining time) for a typical milling operation, thus:

Example 1.19
Determine the standard time for milling the face of a steel block 80 mm wide × 225 mm long, taking one cut 5 mm deep:

a) By peripheral milling using a slab milling cutter 100 mm diameter × 100 mm long, having 10 teeth. Feed 0·2 mm/tooth. Cutting speed 20 m/min.

b) By face milling on centre using a 120 mm diameter face mill, having 14 teeth. Feed 0·4 mm/tooth. Cutting speed 20 m/min.

In each case allow:

5 mm over run.

30 seconds for loading and 30 seconds for unloading.

3 seconds for running table back to starting position by quick power traverse.

2 seconds for approaching work to cutter. 5 seconds for gauging.

15% compensation for relaxation allowance.

Solution
a) Approach $A = \sqrt{5 \times 95} = \sqrt{475} = 22$ (to nearest whole number)

$\therefore$ Travel $L = l + A +$ over run $= 225 + 22 + 5 = 252$

$$\text{Cutter speed} = \frac{1\,000 \times 20}{3\cdot142 \times 100} = 64 \text{ rev/min}$$

Feed/rev $= 10 \times 0\cdot2 = 2$ mm $\therefore f = 2 \times 64 = 128$ mm/min

$\therefore$ Time $T = \dfrac{252}{128} = 1\cdot969$ min $= 118$ seconds

Standard Time:
Load =	30	(by synthesis)
Approach =	2	(estimated)
Mill =	118	(by calculation)
Run back =	3	(by calculation where rate of quick traverse is known)
Unload =	30	(by synthesis)
Gauge =	5	(by synthesis)
Basic time =	188	seconds
15% =	28	
Standard time =	216	seconds

b) Only the machining time is different here.

Approach $A = 60 - \sqrt{3\,600 - 1\,600} = 60 - \sqrt{2\,000} = 15$

Travel $L = 225 + 15 + 5 = 245$

$$\text{Cutter speed} = \frac{1\,000 \times 20}{3\cdot142 \times 120} = 53 \text{ rev/min}$$

$$\text{Feed/rev} = 0.4 \times 14 \qquad = 5.6 \text{ mm}$$

$$\therefore f = 5.6 \times 53 = 297 \text{ mm/min}$$

$$\therefore \text{ Time } T = \frac{245}{297} = 0.825 \text{ min} = 49.5 \text{ seconds}$$

$$\text{Standard time} = (\text{manipulative time} + \text{machining time}) + 15\%$$

$$= (70 + 49.5) + 15\% = 119.5 + 18$$

$$= 137.5 \text{ seconds}$$

In practice the spindle speeds and feeds would be selected from the machine as close to the calculated value as possible. If the batch quantity justified it, pendulum milling could be adopted, i.e., two fixtures mounted on the table, one being unloaded at one end while machining is taking place on the other. Hence the standard time would be less as idle time is reduced.

Exercises 1

1. The following information is known for a component which can be produced upon any of the three machines shown.

Machine	(a) Capstan lathe	(b) Single-spindle automatic	(c) Multi-spindle automatic
Total tooling cost	£20·00	£200	£350
Material cost/piece	£10 for 30	£10 for 30	£10 for 30
Labour cost	£2·50/h	£1·00/h	£0·25/h
Cycle time/piece	5 min	1 min	10s
Setting-up cost	£4·00/h	£4·00/h	£4·00/h
Setting-up time	1h	8h	14h
Machine overheads	£7·50/h	£10·00/h	£11·50/h

Find the three break-even quantities Q_{E1}, Q_{E2} and Q_{E3}, where Q_{E1} is the break-even point between (a) and (b), Q_{E2} is the break-even point between (a) and (c), and Q_{E3} is the break-even point between (b) and (c).

(Ans. $Q_{E1} = 432$, $Q_{E2} = 670$, $Q_{E3} = 1\ 700$)

2. The following information is known for a broaching machine and a milling machine each of which could be tooled up to produce a certain component. Assuming straight line depreciation over 10 years, (i.e., the machine depreciates from its initial cost to zero in equal amounts per year over 10 years), find the break-even quantity Q_E. A 40-hour week is worked over 50 weeks/year.

Machine	Milling machine	Broaching machine
Initial cost	£7 500	£13 500
Tooling cost	£60	£900
Material cost/piece	£0·15	£0·15
Labour cost/hour	£1·60	£1·00
Cycle time/piece	5 min	1 min
Setters rate/hour	£3·20	£3·20
Setting-up time	3h	2h
Machine overheads/hour	£7·50	£12·10

(Ans. $Q_E = 1\ 497$)

3. Three products A, B and C have fixed costs of £25 000, £19 000, £40 000 and break-even quantities of 44 000, 98 000 and 35 000 respectively. Draw a multi product profitgraph. Calculate:
  a) a profit-volume ratio(r) for each product;
  b) the profit (Z) for each product at sales volumes of 60 000, 41 000 and 52 000 respectively;
  c) an equivalent P/V ratio (r) for the system of three products.
    (Ans. (a) $r_A = 0.57, r_B = 0.19, r_C = 1.14$
        (b) $Z_A = £9\,200, Z_B = -£11\,210, Z_C = £19\,280$
        (c) $r_{Equiv.} = 0.66$)

4. A product has fixed costs of £37 500 and variable costs/piece of £2.26. The price/piece of the product is £3.87. Draw a break-even chart. Calculate:
  a) the break-even quantity (Q_E) beyond which a profit is made
  b) the total profit (Z) for a sales volume of 50 000 products.
      (Ans. (a) $Q_E = 23\,292$ (b) $Z = £43\,000$)

5. A project will cost £28 000, and it is estimated to return the following income/year for each of 8 years: £2 000, £4 500, £6 000, £6 500, £7 000, £7 000, £5 000, £4 500. The cost of capital is 16%. Calculate the net present value (N.P.V.) using the D.C.F. method.
        (Ans. N.P.V. $= £6\,153$)

6. On a certain process the fixed costs are equal to £7 500. The material cost is £0·05, labour cost £1·00 for 120 pieces, and the overheads £1·00 for 30 pieces. If the carrying cost is £1·00 for 100 000 pieces, calculate the economic batch quantity and the total cost of the production run. If the labour cost is increased to £1·00 for 96 pieces, what will be the revised total cost?
    (Ans. $Q_B = 27\,400$. $Y_B = £17\,500$. Revised $Y_B = £17\,570$)

7. The following information is known for a machining operation on a component: $n = 0·25$, $C = 150$, tool change time $= 8$ minutes, tool regrind time $= 5$ minutes, machine running cost $= £2·00$ per hour, depreciation of tool per regrind $= £0·125$. Calculate the optimum cutting speed.
  If 1 000 components are required with a machined length of 100 mm at 65 mm diameter using a feed of 0·2 mm/revolution, calculate the time required to machine the batch of 1 000, and the total cost Y_T.
    (Ans. $V_T = 56$ m/min. $Y_T = £80·40$. Time $= 38$h)

8. The material utilization ratios for four components comprising an assembly are 0·1, 0·9, 0·35 and 0·2, and the total costs of production are £25, £56, £0·50 and £5 respectively. Calculate the equivalent material utilization value and by means of a graph show which com-

ponent(s) could most profitably be studied from the material utilization point of view.

(Ans. $M = 0.626$, Component No. 1)

9. Determine the total time to machine a batch of 500 components by face milling on centre, the component being held in a fixture. The face mill is of diameter 200 mm having 20 teeth used at a cutting speed of 30 m/min and a feed of 0·5 mm/tooth. The component is a plain steel block 180 mm long × 100 mm wide. The following should be allowed: 4 mm over run, 1 min for all manipulating and relaxation times per cycle, and 1 hour setting-up time for the batch.

(Ans. 760 minutes)

10. (a) Define (i) *direct cost*, (ii) *indirect cost*, (iii) *fixed cost,* (iv) *variable cost,* and (v) *standard cost.*

(b) Using your own figures show how the total cost per week of running a washing machine in a home laundry can be segregated into (i) direct and indirect costs and (ii) fixed and variable costs.

11. What are *economic batch quantities?* Describe how they can be of help to the production engineer, and show why production at maximum profit is not necessarily the same thing as production at minimum cost.

12. Discuss the economics of metal removal in a production process by single point tools with respect to roughing cuts and finishing cuts.

13. Describe a satisfactory test for determining cutting tool life, and discuss the criteria that can be adopted in order to decide when the tool has come to the end of its useful life.

14. A component required in large quantities can be produced with equal facility by machining from solid or by forging. Discuss and compare the economics of each process taking into account the material utilization ratio for each method.

Further Reading

1) Borthwick C. V. 'Economic Batch Quantities of Machined Items'. *Journal of the Institute of Production Engineers,* Dec. 1970.

2) Eilon S. *Elements of Production Planning and Control.* The Macmillan Co.

3) Evans D. F. and Hemming. *Flexible Budgetary Control and Standard Costing.* Macdonald and Evans.

4) Burbidge J. L. *Standard Batch Control.* Macdonald and Evans.

5) Lissaman A. J. & Martin S. J. *Principles of Engineering Production.* Hodder and Stoughton.

6) Edwards G. A. B. 'Group Technology – Technique, Concept or New Manufacturing System?' *Journal of the Institute of Mechanical Engineers*, Feb. and March 1974.

7) *Glossary of Terms in Work Study*. British Standard 3138: 1969.

8) Currie R. M. *Work Study,* Pitman.

9) *An Engineers Guide to Costing*. Institute of Production Engineers and Institute of Cost and Management Account, 1978.

CHAPTER 2

Ergonomics

2.1 ERGONOMICS AS THE LINK BETWEEN THE ENGINEERING SCIENCES AND HUMAN SCIENCES

THE WORD *ergonomics* is derived from the greek 'ergon' and 'nomos' and means the *laws of work*. Today it is generally called the science of fitting the job to the worker, and concerns the field of the human operator and his working environment. It can therefore be seen that it covers a very wide field, and may be applied at the initial design stage when production lines and workplaces are being laid down; or may be applied to improve the existing layout. Ergonomics is an activity which has developed from *work study,* and the aim is to improve the working environment such that the operators fatigue and strain is reduced, and the efficiency of the manufacturing organisation is increased.

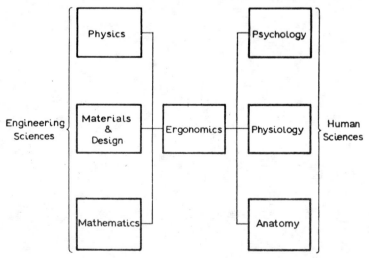

Fig 2.1   Ergonomics linking the Engineering and Human Sciences.

Ergonomics is defined in BS 3138: 1969 as the study of the relation between man and his occupation, equipment and environment, and particularly the application of anatomical, physiological and psychological knowledge to the problems arising therefrom. Do not let the long words put you off; the subject is to do with applying knowledge of the human body and mind to industrial problems. Work study engineers have been doing this for years successfully, but replacing the medical science aspect with intuition and practical common sense. However, in an ergonomics department more specialized knowledge and skills are available than could be possessed by the average work study practitioner.

Figure 2.1 shows ergonomics as the link between the engineering sciences and human sciences.

The engineer studies in the field of engineering technology which is appertaining to the engineering sciences. Now, the activity of ergonomics brings together the two groups of specialists: those who know about machines and processes, and those who know about human capacities. If one studies the old machine in an old, out of date factory, it may sometimes seem that the machine is clumsy and heavy to operate, the controls are out of easy reach, the lighting is poor, the instruments are difficult to read, and the environment is noisy and cold (or too hot). Some might unkindly say that this can also be seen in some so called modern factories. Ergonomics when applied, can show improvement in all the above features. The ergonomist when studying the job with a view to keeping it within human limitations, and in order to make the best use of human abilities, will arrange the things he must consider into three groups, viz.,

a) *Instruments and controls.* This group includes the design of the instruments which the operator must look at or listen to (called the DISPLAY), and the parts of the machine or equipment on which he exerts muscular force so as to change the state of the process or operation (called the CONTROL).

b) *Workplace.* This group includes the work space area around the operator, the working surface and seats.

c) *Working environment.* This group includes conditions of noise, heating and lighting under which the work is to be carried out.

These groups are discussed under the following sections:

Group (a) Sections 2.2 and 2.3
Group (b) Section 2.4
Group (c) Sections 2.5, 2.6 and 2.7.

2.2 ERGONOMICS APPLIED TO INSTRUMENT DESIGN

An instrument is a provider of information, and the information should be displayed to the operator in the simplest manner. For example, a car temperature gauge need only display N for normal temperature, as against a fully calibrated scale which requires the motorist to know what is the normal working temperature for the engine. Display may be grouped into three types depending upon the type of information to be conveyed. These are: (a) Qualitative (b) Quantitative and (c) Representational.

The last type is most uncommon in engineering manufacturing industry, but is more common in the chemical process industries such as a brewery for example.

a) *Qualitative display.* These are used to indicate whether or not a particular function is being carried out, no numerical information being required from the display. They may be visual indicators, such as a red light which indicates that the main drive motor to a machine is running. Or again a red light to indicate that a tool is broken on a transfer line. There is the problem of colour blindness here, and it should be remembered that visual indicators can not only have a different colour but also a different shape and size. A flashing light can be used for very important displays, this being shown to good effect on car direction indicators. There is of course, a convention of colour codes. Red indicates danger to everyone. It may occur to the reader that the start button to switch on a machine tool might well be red (instead of green) in order to indicate that the machine is running when the button is pressed, and is therefore a source of potential danger to the operator.

Auditory or noise indicators, such as a buzzer, have the ability to attract immediate attention, this being a good feature for a warning display.

b) *Quantitative display.* These are used where numerical information may be presented in *analogue* form or *digital* form.

Analogue indicators have a pointer which shows a reading on a scale analogous to the value it represents. A dial test indicator is an example, where one division on the scale may be analagous to 0·01 mm movement of the plunger. Figure 2.2 (a) shows a typical analogue display.

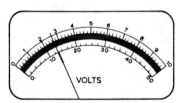
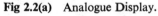
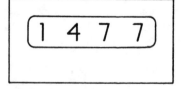

Fig 2.2(a)   Analogue Display.   **Fig 2.2(b)**   Digital Instrument.

The disadvantage here is that a degree of estimation has to be used in order to decide on the position of the pointer between graduations. This type of display may also be qualitative. For example a rev/s meter (rev. counter) may have the scale coloured red beyond a certain value on the scale in order to indicate dangerously high revolutions per second.

Digital indicators show the required information directly as a number. A kilo-meter in a car is an example where the number of kilometres covered can be read directly off the meter as a number. Figure 2.2 (b) shows a digital version of a voltmeter.

The advantage here is that the precise reading to the desired accuracy can be read directly from the meter; however, for a quick approximate reading at a glance, the analogue version gives the best results.

The two types may be combined as shown in Fig 2.3 when a rate indication is required for example (which might possibly be returned to zero as required), in addition to a cumulative quantity.

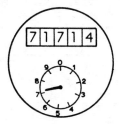

Fig 2.3 Combined Analogue and Digital Instrument.

The important feature is that the scales are legible. White numerals on black background may in many instances prove to be clearer than vice versa. It is interesting to note the design and colour of the numerals on the motorway notice boards, compared to the older form used on most other roads. The type of scale used is important. Figure 2.4 shows two analogue instruments displaying the same information. Instrument B is better than instrument A because:

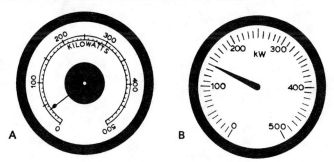

Fig 2.4 Analogue Instruments.

1) Scale B is simpler, bolder and hence more legible.
2) Scale B is longer and divisions clearer because the numerals are inside the scale.
3) Numerals B are upright, hence more legible.

As with colours, there is a well known convention of scales which should be adhered to. These are shown in Fig. 2.5.

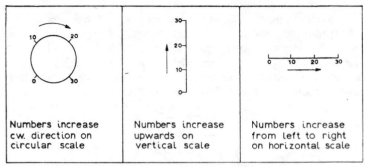

| Numbers increase cw. direction on circular scale | Numbers increase upwards on vertical scale | Numbers increase from left to right on horizontal scale |

Fig 2.5   Instrument Scale Convention.

c) *Representational display*. These are used to give a pictorial diagram or working model of a process. They are ideally suited to large processes because they give an overall display of the process. Examples can be found in modern railway signal boxes where a large display panel shows all the necessary information in a relatively simple manner. Simplicity must be the aim of the designer here, as shown in Fig. 2.6.

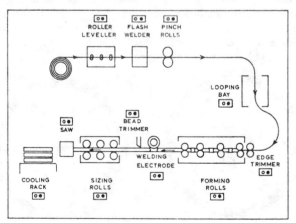

Fig 2.6   Representational Display Showing a Continuous Tube Welding Process.

The display shown would form part of a control panel and gives a visual indication of the state of the process. Light signals, as shown, by each operating point indicate which part of the process is active.

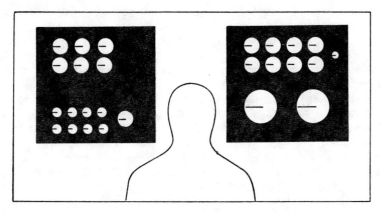

Fig 2.7 Multi-instrument Display.

Multi-instrument display. In the case of a display having many instruments, a logical order or pattern of dials and pointers should be presented to the user. The pointers should all be set so that they are in the same position for the Normal reading of each instrument. See Fig 2.7, where the clear contrast between dial and background should also be noted.

The classic example of this principle is provided in the ergonomic design of the instrument display in the cockpit of a modern airliner.

2.3 ERGONOMICS APPLIED TO MACHINES AND CONTROLS

Levers, knobs, handwheels etc., should be positioned so that the operator can manipulate them with the least change in body position and with the greatest mechanical advantage. The operator should not need to leave his normal working position (which may be sitting or standing) in order to reach a machine control. The controls should be placed comfortably close and in front such that the operator does not need to bend and twist to reach them. The ideal positions may be impossible to attain in the design of a machine because of difference in height and other anatomical dimensions, difference in sex and because of a minority of people being left handed. However, a machine for general use should be designed to suit the average human being, and statistics are available to help to do this. The function of a control should be considered when it is positioned. If fine, delicate adjustment is required then it should be located near the fingers. If heavy force is required then the legs through the medium of the feet can exert a large force. Foot operated controls relieve the hands for other tasks.

Ergonomists with their specialized knowledge have carried out much research into the force, speed and accuracy with which a man or woman can manipulate a control with any limb. Some general conclusions have already been drawn as a result, a few of which are given here:

Force. An operator can employ up to 15% of his maximum strength for an hour or so without rest, and up to 25% of his maximum strength for intermittent short periods all day. (Note — his maximum strength is usually measured over a period of 5 seconds.)

'Jerk' forces which are used to displace tight controls are believed to be twice the magnitude of maximum steady forces. The maximum steady force which can be maintained on a control depends largely upon how the control is positioned relative to the operator.

Speed. This depends upon the shape of the control and again where it is positioned. Hand controls can be operated faster if they are located at, or just below, elbow height.

Accuracy. Depends not only on the design of the control but on the clear presentation of the information to the operator. This was stressed in instrument design in Section 2.2.

Ergonomics principles applied to control design have resulted in certain types of controls being found to be most suitable for certain functions. The following table, which should be read in conjunction with Fig 2.8 gives a summary of the suitability of different types of controls for different purposes.

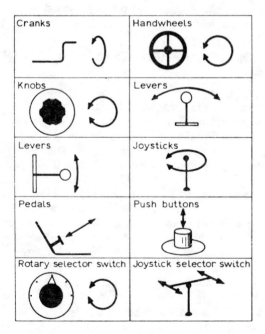

Fig 2.8   Types of Controls for Varying Purposes.

SUITABILITY OF VARIOUS CONTROLS FOR DIFFERENT PURPOSES

Type of Control	Suitability for tasks involving:				
	Speed	*Accuracy*	*Force*	*Range*	*Load or Torque*
Cranks					
Small	Good	Poor	Unsuitable	Good	Up to 4·5 Nm
Large	Poor	Unsuitable	Good	Good	Over 4·5 Nm
Handwheels	Poor	Good	Fair/Poor	Fair	Up to 17 Nm
Knobs	Unsuitable	Fair	Unsuitable	Fair	Up to 2 Nm
Levers					
Horizontal	Good	Poor	Poor	Poor	Up to 110 N*
Vertical (to-from body)	Good	Fair	Short: Poor / Long: Good	Poor	Up to 135 N*
Vertical (across body)	Fair	Fair	Fair	Unsuitable	one hand up to 90 N* / two hands up to 135 N*
Joysticks	Good	Fair	Poor	Poor	20 – 90 N
Pedals	Good	Poor	Good	Unsuitable	130 –900 N depends on leg flexion and body support) (ankle only up to 90 N)
Push buttons	Good	Unsuitable	Unsuitable	Unsuitable	9 N
Rotary Selector Switch	Good	Good	Unsuitable	Unsuitable	Up to 1 Nm
Joystick Selector Switch	Good	Good	Poor	Unsuitable	Up to 135 N

*When operated by a standing operator, depends on body weight.
(Based on Table 4 in *British Productivity Council Seminar on Ergonomics*, Fitting the Job to the Worker by K. F. H. Murrell, 1960.)

In the case of the smaller hand controls, such as knobs or switches which are used for instrument control for example, it will be found that a larger diameter is most suitable for fine, sensitive control. Small diameter knobs can be used for coarse adjustment. Switches operating power are best accompanied by a red indicating light which shows power on. Pointer shapes should be used on knobs which are designed to indicate some value, as shown in Fig 2.9.

Knobs should be distinguishable by shape so that the control can be recognised by feel alone. The knobs shown in Fig 2.10 would not be confused.

Where controls switch on or off, accepted conventions should be rigidly adhered to. With switches, pressing downwards is ON, or with rotary switches, turning clockwise is ON. The writer has experience of two similar milling machines (one vertical and one horizontal), manufactured by the same company, with the same type of switching – on lever on each machine. On one machine the lever was pushed downwards to start the cutter arbor rotating; on the other machine the lever is pulled upwards to start!

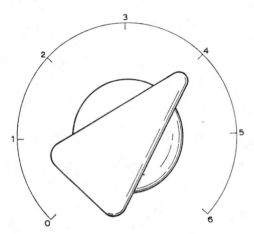

Fig 2.9   Switch Knob.

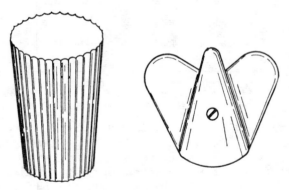

Fig 2.10   Knobs.

Again, where controls are combined with a display, accepted conventions should be rigidly adhered to. These are shown in Fig 2.11.

All the principles with regard to clarity, colour convention etc., outlined in Section 2.2 apply equally well with regard to control layouts on machines. It should be remembered that in a moment of danger when the operator is under stress, he will almost certainly resort to actions which come to him naturally from a lifetime of association.

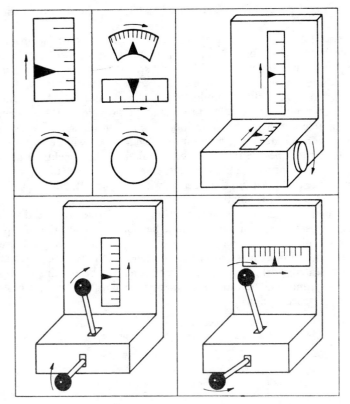

Fig 2.11   Controls Convention.

As is common using work study techniques, a 'mock up' of a console, machine operating position or workplace layout (as outlined in the next section) should be made and tested before a design is finalised.

2.4 ERGONOMICS APPLIED TO THE LAYOUT OF A WORKPLACE

Many of the problems encountered in the ergonomic design of machines and controls will be found in the design of workplace layouts. The workplace is a space in a factory, machine or vehicle which must accommodate an operator or operators, who may be sitting or standing. The efficiency of the operator will depend upon how the workplace is designed. Ideally, a workplace should be custom built for the use of one person whose dimensions are known. For general use, however, a compromise must be made to allow for the varying dimensions of humans. The ergonomist should be knowledgeable, not only in the science of anatomy but also anthropometry, i.e., the measurement of people.

It was stated in the last section that a general purpose machine should be designed for the average person. However, if one considers an operator who may be of either sex, seated or standing at a workplace, there are very few persons with average dimensions for *all* limbs. Therefore, a workplace should be so proportioned that it suits a chosen group of people. Adjustment may be provided (on seat heights for example) to help the situation. If one considers the workplace (driving control area) in a modern motor car which will be required to suit one or two people from a large population, the benefits of adjustment will be appreciated. Ideally, the seat should be adjustable for height, rake and position, the steering wheel for height, and the control levers and pedals for length.

Detailed anthropometric information is available for the different sexes, but simpler information could be obtained once the group of persons liable to use the workplace (or series of workplaces) is known. Figure 2.12 shows suggested critical dimensions for a group of males using a seated workplace. These dimensions can be obtained quickly and easily and will be quite satisfactory for constructing a mock-up of the proposed design.

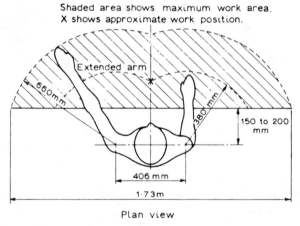

Fig 2.12   Critical Dimensions for Seated Male Operator.

Figure 2.12 shows the left hand covering the maximum working area and the right hand covering the normal working area. These are defined in BS 3138: 1969 as:

a) *Normal working area.* The space within which a seated or standing worker can reach and use tools, materials and equipment when his elbows fall naturally by the side of the body.

b) *Maximum working area.* The space over which a seated or standing worker has to make full length arm movements (i.e., from the shoulder) in order to reach and use tools, materials and equipment.

(NOTE. When the legs are used, similar circumstances apply.)

Assuming the work is some operation requiring equipment, any tools, bins, etc. should be placed within the area shaded so that they can be seen and reached quickly and easily. The body dimensions given in the simple example shown in Fig 2.12 must be the smallest in the group, not the average, in order that all can reach the equipment. If the equipment is numerous, then some tools might have to be placed outside the shaded area thus causing undue reaching for the minority. The position of the seat must be adjustable as shown.

Figure 2.13 shows the situation with respect to bench heights and seat heights.

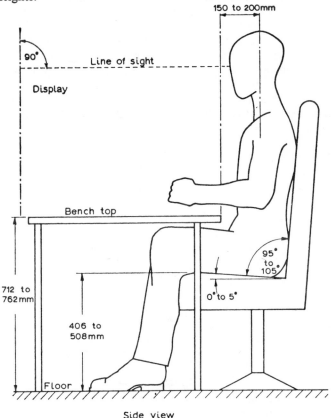

Fig 2.13 Bench and Seat Heights.

In this view the seat should be adjustable for height and rake. It is not usually convenient to have adjustable benches or work tops and the value of 712 mm to 762 mm is probably the best compromise dimension. With fixed bench heights, varying thickness footboards are a help for varying sized operators, particularly where the operator is standing. The design of seats is a specialist activity, and much data is available for dimensions and shapes. A seat should give good support for back, buttocks and thighs as it is a springboard for seated muscular activity. This criterion of support can be seen to conflict with that of comfort if a seat is required to meet both. Yet again, a good compromise usually gives best overall results. Ideally, with a seated workplace, the feet should be firmly on the floor or footboard, body well supported and comfortable and work top on or below elbow height when seated.

If an instrument display forms part of the workplace, then the size of the display should be such as to be easily readable by all. Also the display panel should be at right angles to the line of sight of the operator.

In the case of tote boxes, bins, loose or portable tools, etc., there should be a definite place for their location within the working area. Hence the operator can develop habitual, confident movements when reaching for equipment often without any need for the eyes to direct the hands. The mental effort and strain are less. For the same reason, material and tools used at the workplace should always be located

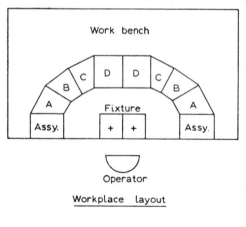

Workplace layout

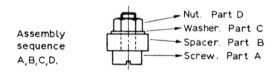

Assembled part

Fig 2.14   Workplace Layout for Assembled Part.

within the working area to permit the best sequence of operations. This is shown in Fig 2.14.

The operation shown consists of assembling four parts A, B, C and D (two assemblies at a time) using both hands. As finished assemblies are placed in chutes, parts A are in the next bins as they are required first for the next assembly.

Where possible clear access should be given around industrial workplaces to allow for adequate supervision and inspection. If workplaces are inside a conveyor system, access must be provided for the operator in case of fire or other emergency. If necessary, this must be done by bridging a conveyor. The Factories Acts stipulate certain regulations which must be observed in respect to 'safe means of access and place of work'. It is clear that if sound ergonomic principles are observed in the design of workplaces, then the operator will be more efficient, less strained and tired and consequently less liable to have an accident. It is difficult to improve a poorly designed layout which has been established for some time.

2.5 NOISE

Many engineering factories are noisy, and given the choice most people would prefer rather less noise than more. Noise can be very annoying, and where concentration is required of an operator in the industrial task he is undertaking, it can be distracting. Therefore, he may work less efficiently and in some instances the risk of accident increases. Noise bothers some people more than others; we all know that a loud musical pop group might please a teenager, and drive a parent into a frenzy of rage. One may become accustomed to familiar noises and eventually take them for granted. In fact one might miss them if they were not present. The unexpected noise is the one which often causes acute annoyance, such as the sudden bang of a press or the piercing whistle of exhausting compressed air. When the bang of a backfire from a car exhaust pipe causes one to jump, the rage is often out of all proportion to the actual volume of noise. Much research is being carried out now on the question of aircraft noises at and near airports, as it is generally believed that they cause a decrease in the health of people who live near the airports. Another factor is that some noises can damage the hearing and cause deafness, and worse still these noises may not necessarily be the loudest or the most annoying. On the grounds of reduced efficiency and damaged hearing then, it is obviously worthwhile to consider how the situation of industrial noise can be improved.

Measurement of Noise

Consider say a horizontal milling process where the nature of the operation often causes undue vibration and noise. The vibration causes a rapid rise and fall in the pressure of the surrounding air. These pressure changes travel through the air in the form of waves and hence may eventually strike the ear of a person. Whether the person hears the sound or not depends upon the amplitude of the sound wave, and how rapidly the source is vibrating.

Two units are used in the measurement of sound. The first is the unit of HERTZ (Hz) which is a measure of the frequency of sound. (Note. Hz is an SI unit which is in common use in noise technology. 5 Hz = 5 c/s.) When this number is small in value the sound will have a low note, and when large in value the sound will be high pitched. A human being will hear sounds between values of 20 and 15 000 Hz. Ultra sonic flaw testing equipment, for example, operates at a frequency to the order of a value of 3 MHz. A sound produced at this frequency is not audible to the human ear. A complex piece of equipment with many different moving parts like a machine tool, will produce different noises each having different frequencies.

The second unit of sound is the DECIBEL (dB) which is a measure of the intensity of sound. The higher the number of decibels, the louder the sound. It can be seen then that frequency measures the pitch of noise and decibel measures the intensity of noise. A very low whisper some few feet away to the order of 1 dB at 1 000 Hz say, would just be detectable to a person with acute hearing. The ears of a capstan lathe operator might be receiving noises to the order of 90 dB at many varying frequencies. One would have to shout to be heard above a noise of this intensity. Decibels are measured on a logarithmic scale such that a tenfold increase in the intensity of noise, whether from 10 100, or from 100 to 1 000, is measured by 10 dB. Therefore, the lathe noise of 90 dB is 10^9 (i.e., 1 000 000 000) times as intense as a whisper of 0 dB. Some industrial noises will be to the order of 140 dB depending upon how close to the offending process one stands.

As stated earlier, the most violent and unexpected noise does not necessarily do the most damage to the hearing, but the less intense noise heard throughout one's industrial life can be most damaging. It has been proved that the ear may become prone to loss of hearing at the particular frequency of this noise, whilst hearing remains un-impaired at many other frequencies. Expert opinion has it that industrial noise first causes hearing loss to occur in the 4 000 Hz region and this damage can spread over a wider range with more prolonged exposure to the noise. To measure industrial noise, meters are available which give a direct reading in decibels, and the frequencies of individual sounds can be isolated and measured using more sophisticated equip-ment.

As a result of research into noise, induced 'deafness' damage risk criteria have been established (Further Reading No. 10) which stipulate, for different periods of exposure, the safe sound pressure levels at different frequencies. The combination of intensity and frequency must always be considered, and expert advice should be taken where any doubt exists.

Precautions against Noise

All noise problems are best tackled at source, and can most economically be taken into account at the design stage. Many machines can be made quieter by improved maintenance procedures, and machines which give rise to vibration should be mounted on properly designed anti-vibration mountings to prevent vibration transmission through the base. Compressed air operated mechanisms are particularly noisy, and exhaust parts should be connected to silencers. When this type of attack has been carried out on noise, often at some considerable expense, it may be found that the level of noise is still intolerable. Then the shielding of the operator from the noise must be considered. It may be possible to enclose all or part of a machine in a soundproof booth. In this case the cost of such an item must be balanced against the cost of the operative's health, and also against the increased efficiency of the working staff which will almost certainly result. Sound proofing materials are available which can be applied to walls and ceilings. These materials are basically of two types which are different in both physical character and design purpose. The first type is insulating material which is in general dense and non-porous, whilst the other type is light and porous and is classified as an absorbent. Both are required in the design of most acoustic enclosures.

In some cases, such as in loom sheds in the textile industry, none of these measures will bring the level of the noise intensity down to an acceptable value. The only alternative left is to fit workers with their own individual ear protectors. Naturally, this will impair speech communication which can in itself lead to accidents.

It is also recommended that people working in high noise level occupations should be given a hearing test at frequent intervals, so that where hearing is affected it can be detected in good time and the operator moved to another job. The level at which damage to hearing can occur differs considerably amongst different individuals.

2.6 HEATING AND VENTILATING

The bodily comfort of the worker depends upon the room temperature in which he works being neither too high nor too low, and ventilation being adequate. As with noise, if these factors are not controlled at a desirable level, then loss of efficiency will occur, health may suffer and rate of accidents may increase. Therefore heating and also ventilating must be considered part of the field of study which the ergonomist must consider and apply. The problem here again is that individualistic human beings are being considered, not cyphers, each with their own particular likes and dislikes. What is a pleasant, warm atmosphere to one man may seem unbearably hot and stuffy to the next. Again what is comfortable for a man is often not comfortable for a woman. Men appear to be more sensibly clothed than women for winter (especially while the fashion for the mini-skirt prevails!), and women appear to be far more comfortable in summer. However, the ergonomists appear to have a range of values fairly well established for our climate, and the Factory Acts have something to say on the subject. It must still be remembered that in the hotter processes, an environment exists which the ordinary person would find impossible to stand for long, but which the operator trained to the work gets used to, and can work in, without apparent detriment to his health. In the pottery industry for example, men work around and near to kilns which operate at very high temperatures, and in the old days of bottle kilns, they worked inside the kilns whilst emptying them, at extraordinarily high temperatures. However it must be admitted that beer is popular amongst pottery workers.

Standards for Industrial Thermal Comfort

There are four factors which the ergonomist will consider when tackling the problem of thermal comfort. These are:
a) Air temperature.
b) Radiant temperature.
c) Air humidity.
d) Rate of air movement.

Of these four, the first is usually classed as the most important. Consider each factor in turn.

a) *Air Temperature.* The Factory Acts state that 'a reasonable temperature must be maintained in each workroom by non-injurious methods. In rooms in which a substantial proportion of the work is done sitting and does not involve serious physical effort, the temperature must not be less than 60° F after the first hour, . . . ' It is generally conceded that 18·3° C (65° F) is a satisfactory standard to aim for although the following values give a more detailed standard:
Workers engaged upon heavy task: 12·8° to 15·6° C (55° to 60° F).
Workers engaged upon light task: 15·6° to 20° C (60° to 68° F).
Workers engaged upon sedentary occupation: 19·4° to 22·8° C (67° to 73° F).

The effect of a process in the workroom such as a furnace, giving off radiant heat, must be taken into account.

In any engineering process where fine machining or fine measurement is being carried out, the standard temperature should be maintained at 20° C (68° F) within one or two degrees. This will necessitate thermostatic control of the room temperature. It will be found that apart from the technical significance of this temperature, most people find it a very comfortable temperature at which to work.

b) *Radiant Temperature*. This is measured using a globe thermometer which records the mean radiant temperature. The desirable standards are as for air temperature. Where radiant heat is being given off from a high temperature process, it is important that the workers are shielded from the heat.

c) *Air Humidity*. This is measured by means of a hygrometer. Absolute humidity is a measure of the actual moisture content in a given weight of air and is not affected by changes in temperature. Relative humidity is the commonly used measurement and is the ratio of the actual moisture content of the air to that which would exist if the air were saturated at the same temperature. It is expressed as a percentage. 65% relative humidity may be considered as the maximum value which should be allowed if personnel comfort is to be maintained. As the temperature falls so humidity may cause discomfort through dryness of the nose and throat, and high humidity may cause discomfort through a feeling of stuffiness and closeness. At ordinary working temperatures humidity has little effect, but at extreme temperatures it does. On a very cold day for example, the relative humidity may fall as low as 20% which can be uncomfortable for some people.

d) *Rate of air movement*. The speed at which air moves through a workroom can effect the comfort of the staff. At 0·5 m/s people will complain of draughts, and at 0·1 m/s the atmosphere will be classed as airless. It has been found that 0·15 m/s is ideal providing that the air temperature and radiant temperature are correct. On the other hand, if the temperatures are well above those recommended then an increase in rate of air movement will be welcomed.

In process industries where dust or fumes are a hazard, these are best extracted at source and exhausted outside. Adequate ventilation is required, and the air should be changed several times an hour.

AIR CONDITIONING

This can mean different things to different people, but today it is generally taken to mean the processing of air to control the factors of temperature, humidity and cleanliness at the required levels. An air conditioning system usually consists of an air circulating system which extracts air from the required rooms, treats it under automatic control and delivers it back to the rooms. In most cases a proportion of outside air will be added to the returned air. Such a system will contain the means of heating (or cooling) the air, humidifying (or de-humidifying) the air, and cleaning the air. Heating is usually done by means of steam or hot water filled coils, or electric elements. Humidifying is usually done by means of water sprays, or steam generating arrangement. Many types of air filtration equipment are available depending upon efficiency of air cleaning required, space available and type of dust in the atmosphere, if any, etc. Apart from personnel comfort, many processes require air conditioning in order to function correctly. Fine measurement was mentioned earlier as one example. Computers for another, will only function correctly in a controlled atmosphere.

Individual heat sources may be used to improve working conditions even if a full air conditioning system is not installed.

HEATING METHODS

These will be designed to give off heat by radiation, convection or a mixture of the two.

Radiators or Convectors

Hot water filled radiators transmit both radiant heat and convected heat and should not be obstructed by equipment. Modern electric convector heaters are very efficient and have the advantage of being portable. Radiant heaters, like electric fires or infra-red wall heaters are a useful means of applying a local source of heat directed at a particular working area. They are particularly useful used outside on loading bays, open sheds, etc., where it is difficult for workers to keep warm.

Floor and Ceiling Heaters

Floor heating systems are popular as they keep feet warm; cold feet being universally unpopular. They are often impracticable for industrial buildings, but are ideal for offices. It is recommended that the floor temperature is not allowed to rise above 25° C (77° F). Skirting heating may be either radiant or convective, the latter often being popular with air conditioning systems where warm air is blown through ducts at skirting board level.

Overhead unit heaters are very popular and have adjustable louvres to enable warm air to be directed where required. They can, however, if badly sited become uncomfortable, giving the people closest to them hot heads. They can also give an uneven distribution of heat throughout a work room. The fan type have the advantage of being dual purpose; they can be used as fans in hot weather, blowing cold air instead of heated air. Radiant ceiling heating is a more efficient but more expensive form of overhead heating. It consists of mounting a battery of radiant heating panels at a suitable overhead height.

Where a heating system only is provided, as against a full system of air conditioning, problems can arise with open windows. They will in some cases be the only means of circulating air in the room if it becomes stuffy. Draughts can arise due to windows being open and can be both unpleasant and injurious to health. Air conditioning, of course, is the complete answer to thermal comfort.

As in other fields in which the ergonomist practises, there are still many questions relating to thermal comfort which remain as yet unanswered. Consider the following:

a) Experiments have been carried out which show that if people enter a centrally heated room and are invited to set the temperature control to suit their own comfort, the majority set the temperature at approx. 23° C (73° F). If the experiment is repeated with a radiant electric fire included as a source of heating, the majority will now choose to set the temperature control at approx. 20° C (68° F). Do people feel warmer because they see heat? Do people feel warmer because they think they are seeing heat, like sitting before an electric fire which has the artificial coal glowing but no bars turned on?

b) It has long been thought that there is some connection between the colour scheme of a room and the comfort of the people working in it. Green is supposed to be an ideal background colour for example, because of its soothing qualities; bright colours are supposed to cheer people up, and so on. But it is now believed in some quarters that a red room is much warmer to work in than a blue room. Why?

c) The effect of noise on people has been discussed earlier, but what about the effect of pleasant noise? Many firms have made it possible for operators to listen to music while they work if they so desire. We would probably all agree that a certain type of music could soothe one, and even make one feel more comfortable. Is it possible that a pleasant noise could make us feel warmer?

These, and many other questions are being investigated by ergonomic researchers.

2.7 LIGHTING

This is the last of the working environment factors to be considered, but is no less important than any of the others. Again, like the others, if light is controlled at a desirable level, satisfactory and comfortable to the working staff, efficiency will be high. Hence, productivity will be higher as a result, health will be improved and less accidents should occur. The difficulty in the field of ergonomics is to prove that these desirable results do in fact occur. Controlled experiments to prove that improved lighting (or reduced noise, or better heating) increases production are almost impossible to carry out on the factory floor. The very act of experimenting with a chosen group of workers alters the system and their attitude to their task. The reader might be interested to read of the Hawthorne experiments carried out on a chosen group of workers in an industrial unit in America, where the effect of changing environmental conditions was studied. (Further Reading No. 8.) The only firm conclusion drawn was that the classic method of experimentation of isolating all parameters except one, was not applicable to an industrial engineering situation.

However, from the evidence available it does appear that generally management is convinced that improved lighting does increase working efficiency. The cost of lighting is small in relation to many other costs, and on the average is to the order of 2% of total wages cost. Most lighting installations can be assumed to have a life of 10 years, and costing studies have shown that the cost of new lighting could possibly have been paid off in one year, if the increased output resulted from the improved lighting alone.

Measurement and Amount of Light

The unit of light (Luminous flux) is the lumen, which can be defined as the amount of light emitted by a point source having a uniform intensity of one candela. The minimum amount of light required for reading, writing, etc., is 10 lumens/m^2, i.e., the light given off by ten (international) candles at a distance of one metre from the work. Note here that the SI unit of illumination is called the *lux* which is an illumination of one lumen per square metre. Much more light than this is needed for reading without strain, of course, but the amount varies depending upon the person's sharpness of vision and the environment in which the reading is being done.

Various bodies have issued codes of practice. The Factories Act states that 'there must be sufficient and suitable lighting in every part of the factory in which persons are working or passing', and requires a minimum of 65 lux. The Illuminating Engineering Society of Great Britain publishes a Code of Good Interior Lighting Practice (1961) in which it is proposed that a lighting level of 160 lux should be regarded as minimum in all work places as a general amenity. The Code also gives specific recommendations a few of which are given here.

Visual Task	*Recommended Illumination*
	(lux)
Rough assembly and inspection work	160
Very fine assembly and inspection work	1 600
Weaving light cloth	320
Weaving dark cloth	750
Sheet metal work	220
Planing wood at machine or bench	220

It is considered by many experts to be uneconomic to provide less than the IES standards, and could be profitable to double them or more. An interesting industrial experiment was carried out in Germany in connection with this. (Further Reading No. 9.) This was undertaken by the Wiesbaden Research Institute of Work Psychology and Personnel Affairs at a handbag manufacturers. The output from 12 workers carrying out a leather stamping operation was measured for two 'control' years at a lighting level of 380 lux. This was compared with the output for two 'test' years at a lighting level of 1 100 lux. The average increase in production was 7·6%. This value represented 13 times the annual total cost of the lighting installation. Even considering the qualifications about experimental results mentioned earlier in this section, these results seem significant.

The level of illumination at any workplace can be easily and cheaply measured using a correctly calibrated light meter. The level can be compared with the IES Code, and the result may also prove to be illuminating!

Types of Electric Lamps

Three main types of lamps are available for industrial use. These are (a) Tungsten filament, (b) Tubular fluorescent, and (c) Discharge. These have all been improved to a remarkable extent in recent years due to intensive technical development.

a) *Tungsten filament lamps.* These produce light when the tungsten wire inside the glass bulb is heated by the passage of electric current to incandescence. They are cheap with a colour rendering which is acceptable for many purposes. Recently, tungsten halogen lamps have become available where smaller bulbs are used in which the rate of evaporation of the tungsten filament is reduced. This in turn gives twice the life of an equivalent filament lamp at a much higher light output. Tungsten filament lamps are simple in operation in that they do not need control gear such as chokes and capacitors as do fluorescent tubes and discharge lamps.

b) *Fluorescent tubes.* The tube in which the discharge takes place is coated on the inside with a fluorescent material. They are far more efficient (ie., luminous efficiency; measured in lumens per watt) than filament lamps but the wattage range is much lower. The newest tubes now available have better colour rendering properties than the common white tube.

c) *Discharge lamps.* Light is produced in a discharge lamp by striking an arc in a metallic vapour atmosphere. The colour of the light emitted depends upon the metal used, sodium and mercury being the most commonly utilized elements. These lamps have a high luminous efficiency, particularly the modern, high-pressure sodium lamps. All discharge lamps are available in a convenient bulb shape, in a wide range of sizes (expressed in watts).

The features of these lamps can be compared in the table below.

Type of lamp	Wattage range	Efficiency (lumens per watt)	Rated life (hours)
Tungsten filament	40–1 000	13	1 000
Tungsten halogen	300–2 000	21	2 000
Flourescent tube (white)	6–125	76	8 000
High pressure mercury discharge	50–1 000	56	7 500
Low pressure sodium discharge	35–180	200	7 500
High pressure sodium discharge	70–1 000	117	7 500

(Based on Table 2 in 'Lighter costs with modern lighting systems' by J. D. Lovatt. See Further Reading No. 7)

Glare

There may be ample light for comfortable working in a work area, but the operator may suffer visual discomfort because of glare. Take the example of an engineering inspector working at a marking-off table which has a good reflective surface. Glare from a badly positioned light source can be reflected from the table top reducing his visibility to such an extent that he cannot clearly see the graduations upon his measuring instrument. He may be unaware of this happening, but only knows that the graduations are not very clear. The degree of glare depends upon the brightness and area of light source creating the glare, their position and the brightness of the surroundings.

The D.S.I.R. Building Research Station in collaboration with the I.E.S. have managed after much research to quantify glare. The amount of glare can now be expressed precisely by means of a Glare Index in terms of the brightness and position of the light sources causing the glare. Some examples of limiting Glare Index are given here from the I.E.S. Code.

Visual Task	*Limiting Glare Index*
Rough assembly and inspection work	28
Very fine assembly and inspection work	19
Weaving light cloth	19
Weaving dark cloth	19
Sheet metal work	25
Planing wood at machine or bench	22

Good Lighting Practice

The field of lighting, like those of noise and heating, is a highly specialized activity but some general principles of good ergonomic practice will be given here. They can be placed under the headings of (i) Daylight, (ii) Artificial Light, and (iii) Mixed Daylight and Artificial Light.

i) *Daylight*. The diversity of illumination should not in general exceed 1·5 to 1.

Lighting systems which allow daylight from one direction only, such as the north light or saw tooth roof, should be avoided.

Factory services should not obstruct the light. This means that underfloor service passages might have to be used.

Floors and ceilings should be light in colour to reflect the light.

An illumination level of 270 lux between 8 am. and 5 pm. should be aimed at for industrial purposes. This can be achieved by using a glazed area of $\frac{1}{10}$ to $\frac{1}{5}$ of the floor area.

ii) *Artificial Light*. There should be a general level of lighting of at least 220 lux with local lighting in addition where necessary, so designed that no glare is caused in the working area.

Tubular fluorescent lamps are best used at low mounting heights.

Mercury discharge lamps (colour corrected) are best used at high mounting heights.

The spacing between light fittings must be related to height for optimum results, and for each type of fitting there is a spacing/height ratio.

The workspace should look more bright and colourful than the general surroundings.

Light fittings should be regularly maintained and cleaned.

The lighting scheme, as with other ergonomic considerations, should be considered as part of the whole working environment.

Most people, given a free choice, will choose a level of lighting to the order of 1 000 lux and more; this is likely to be the minimum level specified in future years.

iii) *Mixed Daylight and Artificial Light.* Artificial light is cheaper than daylight. In multi-storey buildings, daylight from the side windows has little effect on the general level of lighting.

In single-storey buildings, of which type most exist in industry, one can often choose illumination by either source of light, or a mixture. Most people seem to prefer a mixture with daylight more dominant than artificial light, but sufficient of the latter to give a similar level of light at night.

The colour of artificial light should be matched to daylight such that one cannot notice the difference between the two.

Exercises 2

1. What is an Ergonomist? What special skills does he require and how do you think these skills can be acquired?

2. Define a (i) qualitative (ii) quantitative and (iii) a representational instrument display respectively. Give one example of the application of each type.

3. Sketch the outline of a suggested instrument display for a family saloon motor car which should include the means of indicating (a) Road speed in km/h (b) Quantity of petrol contained in the tank (c) Engine running temperature (d) Oil pressure (e) Ignition working (f) Total distance covered (g) Battery discharge in amps.

4. Examine the controls of a machine tool in the College workshop. Constructively criticize the layout of the controls from the point of view of accessibility for the operator, safety, force required to operate them, colour, pictorial representation which indicates on a panel close to the control how it should be operated (if any), and the direction of operation of the control (check if the accepted conventions are observed).

Suggest any alterations you think would improve the layout.

5. Assume you have to design a table and chair on which a simple assembly operation is to be carried out by every member of your class. Obtain the appropriate limb dimensions for all the class, and hence arrive at the most suitable bench height, seat height and working areas for the table.

6. How is noise measured? Explain the effect of varying the frequency and the intensity of sound.

7. Using a thermometer measure the air temperature in various rooms and at various points in the corridor and hall and also outside the building. Plot the temperature gradients you obtain. Observe the activities being carried out in the areas surveyed and check the methods used for heating and ventilating. Comment.

8. Carry out a similar survey to the above for lighting using a light meter as the means of measurement.

9. 13 amp. electric plugs are to be assembled by hand. Mock-up this process on a table fixing the position of the boxes holding the various component parts. By varying the position and amount of light on the workplace, ascertain if there is any change in the 'operators' production rate as a result. Was there any glare from the lights at any time in the test, and if so how did it occur?

Further Reading

1) 'The Factories Act 1961. A Short Guide.' *H.M.S.O.*

2) 'Ergonomics in the Factory: Fitting the Job to the Worker.' *British Productivity Council.*

3) 'Glossary of Terms in Work Study.' British Standard 3138: 1969.

4) 'System for the Direction of Rotation of Machine Tool Handwheels and Levers,' British Standard 754.

5) Murrell K. F. H. *Ergonomics.* Chapman and Hall, London.

6) Eaton-Williams R.H. 'Air Conditioning in Production.' *Journal of the Institute of Production Engineers*, Oct. 1965.

7) Lovett J. D. 'Lighter costs with modern lighting systems' *Journal of the Institute of Mechanical Engineers*, Nov. 1976.

8) Roethlisberger F. J. and Dickson W. J. *Management and the Worker.*

9) Stenzel A. G. 'Erfahringen Mit 1 000 1x In Einer Lederwaren-fabrik.' Lichttechnik, Berlin, January 1965.

10) Burns W. 'Noise as an Environmental Factor in Industry.' Transactions of the Assoc. of Industrial Medical Officers, 14 July 1964.

11) Burns W. and Littler T. S. *Modern Trends in Occupational Health.* Butterworth & Co.

12) Metcalfe T. and Metcalfe J. 'The Reduction of Workshop Noise.' *Journal of the Institute of Production Engineers*, Dec. 1974.

13) Booth M. L. 'Materials and Methods for noise reduction.' *Journal of the Institute of Mechanical Engineers.* Dec 1976.

14) Corlett E. N. 'The factory and its people.' *Journal of the Institute of Production Engineers*, Oct. 1974.

CHAPTER 3

Jigs and Fixtures

3.1 DIFFERENCE BETWEEN JIGS AND FIXTURES

IN ENGINEERING there is often more than one way of manufacturing a component, and cost is often the criterion of choice of method as discussed in Chapter 1. If a component is required in large quantities, then clearly a method which is suitable for producing one-off (such as marking out, setting on machine, clamping to machine table, etc.) would not be suitable for economic reasons. A faster and more profitable method requires some device on which the component(s) can quickly be positioned in the correct relationship to the cutting tool(s) and quickly clamped before machining takes place. Such a device is known as a jig or fixture.

A *jig* is a device usually made of metal which locates and holds the workpiece(s) in a positive manner and also guides the cutting tool(s) such that it is in the correct relationship to the work when machining commences. It is usually necessary for the work to be held in the jig by clamping. The jig is not fixed to the machine table by clamping but is held by hand. Jigs are used for quantity drilling, reaming and tapping for example.

A *fixture* is a device similar to a jig but as the name implies is fixed to the machine bed by clamping in such a position that the work is in the correct relationship to the cutter. A further difference is that the cutter is not guided into position ready for machining to commence. A setting gauge is often provided to enable the initial setting of work to the cutter to be quickly and easily accomplished before production begins. Fixtures are used for quantity milling, turning and grinding for example.

These definitions are not always precisely applied in engineering, and the terms jig or fixture are often used quite loosely.

The practice of designing and making jigs and fixtures has been in wide use for many years, and the numbers of these devices successfully used is legion. We are therefore considering a vast subject in which a great deal of personal preference, variety, inventiveness and even engineering dogma may be found. Therefore the student may be puzzled at first sight by the apparent lack of uniformity. However, certain sound design principles are now well established and will shortly be discussed, and also much standardization has been applied in this field of production engineering leading to cheaper production methods.

3.2 PRINCIPLES OF LOCATION

Kinematics is the branch of mechanics relating to problems of motion and position, and kinematic principles can be applied in considering the location of work in a jig or fixture. Figure 3.1 shows a body in space.

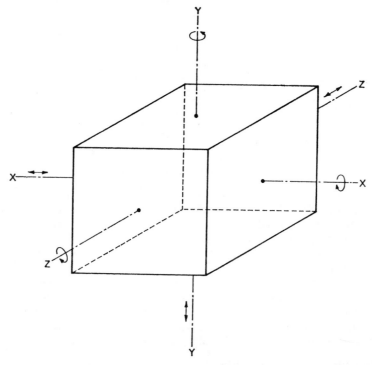

Fig 3.1 Body in Space.

It can be seen that the body has six degrees of freedom. It can rotate about, or have linear movement along each of three axes, XX, YY or ZZ. It is not constrained or prevented from moving in any direction. When located in a jig a workpiece must be constrained from moving in any direction. This can be done by six locations in the case of the body shown in Fig 3.1, this being known as the six point location principle. It is illustrated in Fig. 3.2.

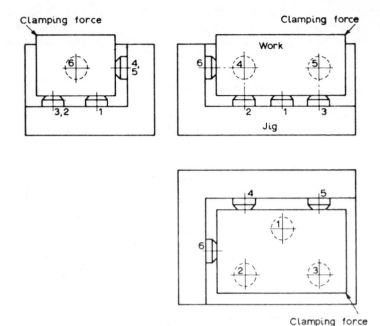

Fig 3.2 Six point Location Principle.

The base of the component is resting on three location pegs which is the minimum number of points upon which it will firmly seat. When the closing force is applied (by means of a clamp say) all the six degrees of freedom have been removed. If more than six points are used the additional points will be surplus and unnecessary and would therefore be redundant constraints.

Depending upon the shape of the component six location pegs may not be required in practice and may be replaced by other devices. However the same basic principle prevails. Three examples of alternative means of location are given next, each of which has removed the six degrees of freedom without having redundant constraints.

Location upon two Plugs

The smaller hardened and ground plug has flats machined on each side to allow for slight variations in centre distance of the two holes in the lever. Two completely round location plugs would allow much less variation in the hole centres, thus providing a very precise form of location.

Location in two Vees

This is a common form of location when the shape of the component will allow it to be used. With one vee sliding as shown it can also be used for clamping as well as locating. A combination of both location methods shown in Figs 3.3 and 3.4 can be used by locating the component upon a round plug at one end and in a sliding vee at the other end.

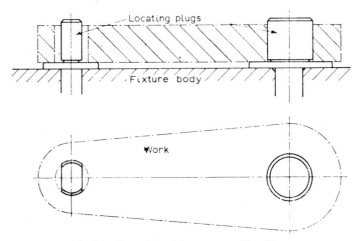

Fig 3.3 Location of Lever upon Two Plugs.

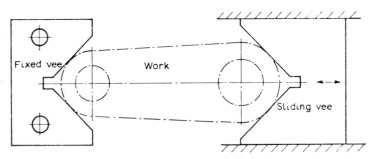

Fig 3.4 Location in Two Vees.

Location upon a Plug or in a Spigot Recess

Depending upon the sequence of machining operations a component may have to be located internally upon a plug [Fig 3.5(a)] or located externally in a recess [Fig 3.5(b)].

Plugs should not be screwed into the fixture body as a threaded hole will not allow the plug to be positioned in the fixture with sufficient accuracy.

Spigot recesses may be cylindrical or of some particular profile to suit the shape of the component.

Design Features to consider with respect to Location

a) Location influences the accuracy and quality of a component. Location should be designed to *kinematic principles* thus reducing the

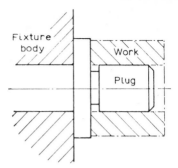

Fig 3.5(a)   Location upon a Plug.

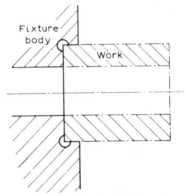

Fig 3.5(b)   Location in a Spigot Recess.

six degrees of freedom to zero with no redundant location features.

b) With first operation work upon rough unmachined surfaces use three point location where possible, and adjustable or expanding locators to allow for large variations in size.

c) Locate from the same machined surface (datum) for as many operations as possible in order to reduce the possibility of error.

d) Make sure the location is 'foolproof', i.e., the component can only be loaded into the fixture in the correct position.

e) Location features should not be swarf traps and should have clearance provided where necessary to clear machining burrs.

f) Consider the operator's safety and the manipulative difficulties in loading the component into the fixture. Retractable location pins may be necessary.

3.3 PRINCIPLES OF CLAMPING

As with location, clamping will influence the accuracy and quality of the component and will further influence substantially the speed and efficiency of the operation being carried out on the component. Many types of clamps and clamping methods have been standardized and are suitable for the great majority of engineering operations. The main principle to observe when designing any clamping arrangement is that the fixture, workpiece and location features should not be distorted or strained. This means that clamping forces should not be excessive but sufficient to hold the work rigidly and should be applied at points where the work has the support of the solid metal of the fixture body. Three examples of alternative means of clamping are given next, chosen from a great variety of standard examples.

Bridge Clamp

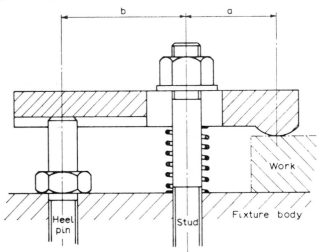

'a' should not be greater than 'b'

Mechanical advantage $\quad MA = \dfrac{a+b}{b}$

Fig 3.6   Bridge Clamp.

As the nut is unscrewed the spring pushes the clamp upwards. The clamp has a longitudinal slot so that it can be pushed clear of the work. To speed up the clamping operation the hexagonal nut may be replaced by a threaded handle, or a quick action locking cam.

Two-way Clamp

This clamping system enables a two-way clamping action to be obtained from one nut or threaded handle. Clamping force is applied to the top and one side of the work piece. The clamp has a quick release action. As the nut is released, the hinged stud can be swung clear from the slot in the end of the top hinged clamp. More elaborate versions of this and other multi-way clamps have been in use for many years.

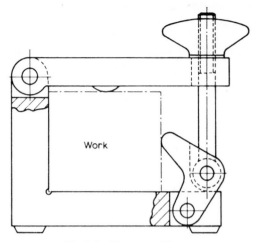

Fig 3.7   Two way Clamp.

Wedge Operated Clamp

$$\phi = \frac{360m}{\pi D \tan \alpha}$$

$$f = \frac{FD \tan \alpha}{2R}$$

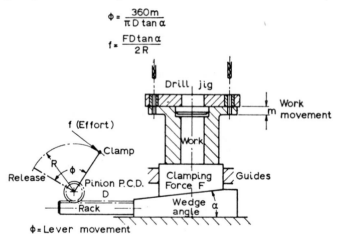

Fig 3.8   Wedge operated Clamp.

$$MA = \frac{F}{f}$$

It can be shown that $f = \dfrac{FD \tan \alpha}{2R}$ assuming 100% efficiency for the mechanism.

Horizontal linear movement of the wedge clamps the workpiece vertically. Alternative manually operated devices for the wedge are a screw or cam. A further alternative is that the wedge could be operated by pneumatic cylinder or hydraulic cylinder. This then leads to the possibility of automatic clamping of the workpiece as part of a fully automatic machining cycle. This type of arrangement will be discussed further in later chapters. Where air or fluid power is being used for actuating clamps, safety arrangements must be made to protect the operator from trapping his fingers or hand.

Design Features to consider with respect to Clamping

a) Clamping influences the accuracy and quality of a component. Clamps should be applied to the component where it is rigid and well supported.

b) Clamping forces should be controlled such that they are not great enough to distort the location, work, fixture, or the clamps themselves. Note that the magnitude of cutting forces to be resisted varies with different machining operations.

c) Clamps should be quick acting and as simple as possible. Some quick acting devices are expensive and the costs should be considered to see that such devices are justified.

d) Ensure that the clamps are well clear of cutting tools, and that adequate clearance is available with the clamps released to load and unload the work in safety. Ejectors may be necessary for ease of unloading.

e) Sound ergonomic principles should be applied when designing the clamping system to be used by an operator. (See Chapter 2.)

3.4 DESIGN FEATURES OF JIGS AND FIXTURES

Many considerations, including technical and economic, will influence the jig and tool designer before the final fixture design is evolved. The magnitude and importance of these considerations will of course depend upon how simple or elaborate the production process is to be. Despite standardization nearly every fixture or jig is a unique design and creation, and the reader is urged to observe that the truly functional fixture or tool can also be a thing of engineering beauty. However, whatever the process or quantities required, certain features are of major importance in any fixture design. These are:

a) Method of locating the component(s) in the fixture.

b) Method of clamping the component(s) in the fixture.

c) Method of positioning the tool relative to the component.

d) Method of positioning the fixture relative to the machine. This is not required with a jig which is not clamped in some fixed position upon a machine table, but even so some form of stop may be provided against which the jig may be held to give a quick initial positioning.

e) Safety aspects to protect the operator whilst using the fixture.

A complete text book is required to analyse all the great variety of these major features as applied in various machining processes, but one example of a jig or fixture for the common processes such as drilling, turning, etc., will be given in order to illustrate good practice.

3.5 DRILL JIG

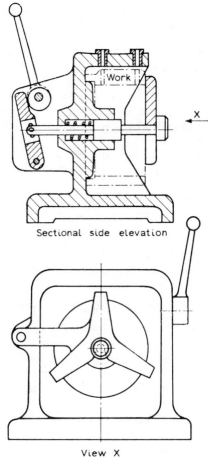

Sectional side elevation

View X

Fig 3.9   Drill Jig.

The following design features should be noted.

a) Location is from a previously machined bore in the component on to a hardened and ground male spigot.

b) Clamping is against the spigot shoulder with a spider clamp plate, actuated by a cam operated plunger. The method adopted will not cause distortion. When the cam lever is pulled to the release position the spider clamp is easily and quickly removed facilitating rapid unloading of workpiece.

c) The cutting tool (twist drill) is positioned relative to the work by means of hardened and ground drill bushes which act as guides for the drill. Proportions for these bushes are given by British Standard Specification 1098. Thè gap between bush and work is large enough to avoid being a swarf trap, but not too large to give the drill inadequate support as it enters the work.

d) No method of positioning the jig relative to the drilling machine table is necessary as the jig is held by hand.

e) A guard should be provided around the drill. Adequate clearance is provided so that the swarf can easily fall clear through the jig body and spider clamp, and no swarf traps are present. Hence the danger of the drill breaking and causing injury is greatly reduced. The jig has no sharp corners.

Further features that form part of good drill jig design practice are:

f) The jig which is handled frequently during use is of light construction consistent with rigidity. The jig body can be a casting as shown or alternatively it can be fabricated from screwed and dowelled steel plate.

g) The axial cutting force is not taken directly by the clamps as is desirable. This should always be so although it is sometimes impossible to observe this general principle.

h) The base of the jig is so designed that the feet cannot fall into the drilling machine table tee slots.

i) Foolproofing pins so arranged that the component cannot be placed in the jig other than in the correct position.

3.6 MILLING FIXTURE

The following design features should be noted.

a) No precise location required as this is the first operation to be carried out on a casting. The component flange rests upon four hardened and ground pins (four being necessary in this case for stability) which have domed tops giving point contact on the uneven casting surface. With great variation of casting size, the pins would have to be made adjustable and capable of being locked in the required position.

b) The clamping arrangement has to be below the top surface of the component which is to be face or slab milled. The side bridge clamp shown with its serrated edge pulls the component firmly down on to the pegs and against the opposite serrated jaw. The clamp is of the equalizing type which allows for considerable variation in the casting size. This fixture, as with many milling fixtures, could be designed to hold more than one component in tandem by multiplying the number of clamping arrangements.

Milling cutting forces are high in magnitude, hence the clamps must be of robust design. Cam arrangements should be avoided as they can

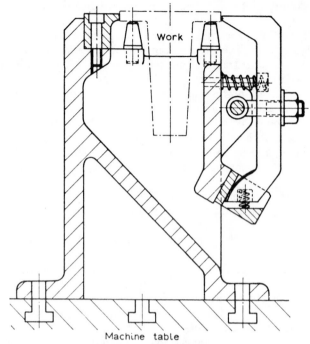

Work

Fig 3.10   Milling Fixture.

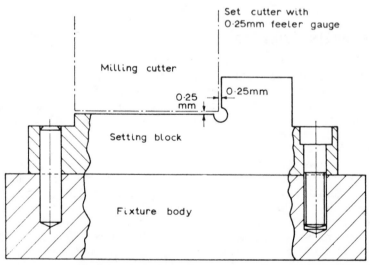

Set cutter with
0·25mm feeler gauge

Milling cutter

0·25mm

0·25
mm

Setting block

Fixture body

Fig 3.11   Setting Block for Milling Fixture.

be loosened during cutting and hexagon nuts and spanners are preferable.

c) Precise positioning of the milling cutter relative to the work is not necessary, but when required a setting block screwed and dowelled in the correct position on the fixture body can be used in conjunction with a feeler gauge. This is shown in Fig 3.11.

d) The fixture would require bolting to the machine table while machining takes place. Again precise positioning is not necessary in this case, but where it is necessary, tenon blocks screwed to the underside of the fixture can be used to locate the fixture on the machine table in the correct position. This ensures that the fixture is in line with the axis of the machine table. This is shown in Fig 3.12.

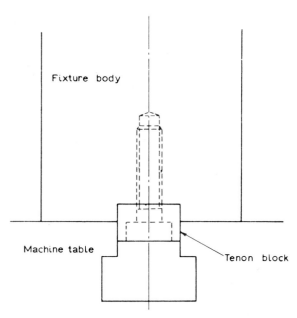

Fig 3.12 Tenon Location between Fixture and Machine Table.

e) Milling can be a very dangerous machining operation. The machine should never be started up without a guard around the cutter, and a limit switch could be arranged to ensure that this is always so. The release points for the clamps should always be well clear of the cutter.

Further features that form part of good milling fixture design practice are:

f) The fixture is of heavy, rigid construction to withstand the high milling cutting forces. A cast iron casting is chosen for the fixture body for this reason and also because it has the property of damping the vibration present during the milling operation.

g) With up cut milling technique using a slab mill there is a component of the cutting forces tending to lift the component up out of the fixture. With the wedge shaped serrated clamp and jaw used this is successfully counteracted.

h) The work is set against a solid face to resist the considerable feed force of the milling operation. This prevents any possibility of the work being pushed out of the fixture. This is shown in Fig 3.13.

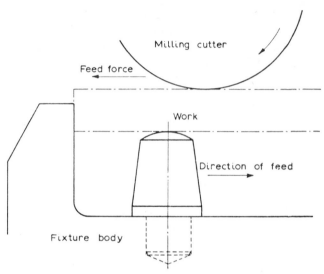

Fig 3.13   Feed Force directed against Solid Metal.

3.7 TURNING FIXTURE

Figure 3.14 shows a sketch of a turning fixture and the component which it is designed to accommodate.

The following design features should be noted.

a) The component is located in a horseshoe slot (as opposed to a plain bore) for ease of loading and is also located flat against the location spigot to ensure that the bore is square to the flat faces of the fork.

b) Clamping is by means of a bridge or latch clamp through the previously machined slot. The stud is pivoted to allow quick release after untightening nut. The clamp is so arranged that it will only fit through the slot if the component is correctly aligned.

c) The location features of the fixture are so arranged that the drill and reamer will be correctly positioned relative to the work when they are fitted in the capstan lathe turret.

d) The fixture is positioned relative to the machine by means of a location spigot on the back of the fixture body which fits the recess in the spindle flange on the machine.

e) All parts of the fixture are within the outside diameter of the fixture body. Integral balance weights ensure that there are no out of balance forces which could cause danger (particularly at high speeds) to the operator. A guard must be provided which completely covers the fixture when it is revolving. The question of balancing the fixture when loaded is important and can be considered as a further feature apart from that of safety.

f) A balance weight(s) is required to ensure the geometric accuracy of the hole when bored. The bored hole will not be truly round if machined with the fixture badly out of balance.

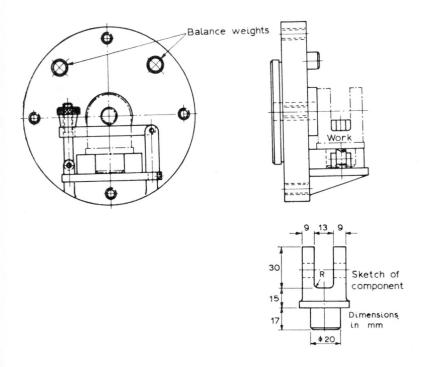

Fig 3.14 Turning Fixture.

3.8 GRINDING FIXTURE

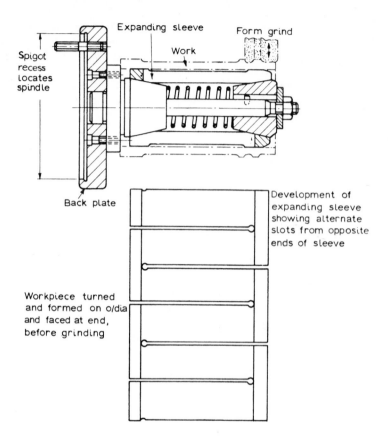

Fig 3.15   Grinding Fixture.

There are no design features peculiar to grinding fixtures that have not already been covered. Cylindrical grinding fixtures are similar in principle to turning fixtures, and surface grinding fixtures are similar in principle to milling fixtures. Neither need be as robust since grinding cutting forces are comparatively light. Much simple work can easily be located and gripped on magnetic chucks with little risk of distortion.

In view of the above remarks, an expanding mandrel has been chosen as an example of a grinding fixture to add to the variety of the examples. The work is being form ground on a cylindrical grinding machine, and the gripping power of the expanding sleeve is adequate to resist the grinding forces. The design of the sleeve is such that it imparts a grip on all the surface area of the previously machined bore.

It is particularly important with a finishing process such as grinding that the component does not move after being released from the clamps hence destroying the accuracy. For the same reason of maintaining accuracy, spindle spigot fits must be precise, balancing where required must be precise, and rigidity is important to stop vibration during grinding.

3.9 HORIZONTAL BORING FIXTURE

Note that this type of holding device for boring, commonly called a fixture, is in effect a jig by definition, as the cutting tools may be guided from the jig. The horizontal boring process is ideal for components which would be difficult to machine on a lathe because of their size and shape.

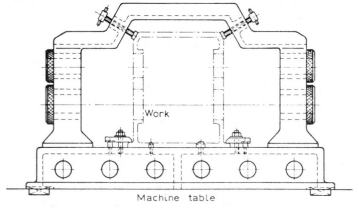

Fig 3.16   Horizontal Boring Fixture.

The following design features should be noted.
a) The component is located upon pins through holes in the base all of which were machined in previous operations.
b) Clamping is by simple bridge clamps, access to which is provided through the open jig body. Steadying screws are provided overhead as the height of the component tends to make it unstable. Clamping need not be heavy as the cutting forces during boring are not excessive.
c) Positioning of the boring tools relative to the work involves designing two features into the jig. First, close (running) fit steady bushes are provided at each end of the jig into which the boring bar is fitted. Secondly, a setting gauge is provided to ensure that the boring tool can easily be set to machine the correct diameter bore. See Fig 3.17.

If a roughing and finishing tool is incorporated in the same bar then the gauge can be as shown.
d) The jig is positioned relative to the machine by fitting tenons to the underside of the jig base as described earlier under Section 3.6.

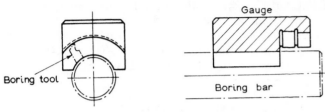

Fig 3.17   Setting Gauge for Boring Tools.

e) Horizontal boring is a comparatively safe machining operation, but quick fitting guards should be used to cover the revolving tools during cutting.

Further design features to note are:

f) Adequate clearance should be provided at each end between the jig and the work to allow easy setting and locking of the tool bits.

g) The boring bar should be easy to fit and release from the machine spindle nose as this has to be done very time a component is bored.

h) Several tools can be accommodated in the boring bar depending upon the operations required. These tools include roughing and finishing boring tools, shell reamers, facing tools and chamfering tools.

3.10 WELDING FIXTURE

Welding fixtures can be classed as assembly and welding fixtures. A typical operation is that of locating and clamping several components in a fixture ready for welding into a fabricated assembly. The process may be gas or electric arc welding, spot or seam resistance welding, or even brazing or soldering. A fixture in which parts are assembled ready for electric arc welding is shown in Fig 3.18.

The location and clamping principles described earlier for machining operations should be adhered to, but there are some design features applicable only to welding fixtures which should be considered. These are:

a) Expansion will take place due to the great local heat generated by the welding process. Should the component distort or buckle as a result, the locations should be so designed that the welded assembly does not lock in the fixture hence preventing its removal.

b) The fixture should be rigid and stable for the same reason. In aircraft assembly work, fixtures of this type may be of enormous size, and tubular fabricated construction is often preferred for these fixtures giving great strength with lightness.

c) Clamps should be arranged in such a way that they do not distort the component. Heavy clamping pressure is often unnecessary and finger screw operated clamps or toggle clamps are frequently used. On car body assembly lines many examples of toggle clamping can be seen, much of it pneumatically operated.

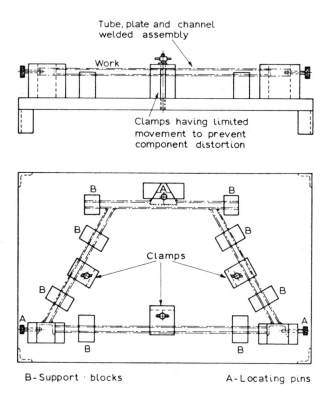

Fig 3.18   Welding Fixture.

d) Weld spatter can be a problem where electric arc welding is used particularly if it adheres to clamp screws. Where possible clamps should be kept clear of the welding zone, or should be shielded against the process.

e) The location and clamping arrangement should not be so eaborate, that the welding zone is inaccessible to the welding operator. Accessibility and simplicity are important factors.

f) In small, spot welded assemblies, a simple nutcracker type of jig is often adequate. This can be held in the hand, the work being presented under the electrodes by the operator, and the position of the weld 'slug' being judged by eye. If the spacing of the slugs is important then an indexing fixture can be used with a single pair of electrodes. See Fig 3.19.

Alternatively, a multi-electrode machine could be used if economically justified, where many welds are made simultaneously. Examples of this can be seen on car body assembly lines where a large pressing such as a floor panel can be welded very quickly to the body.

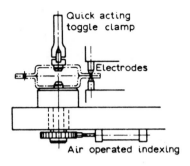

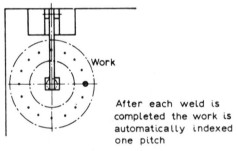

Fig 3.19   Spotwelding Indexing Fixture.

3.11 DESIGN STUDY PRINCIPLES

Before any fixture design is finalized all of the factors mentioned previously, both economic and technical, have to be considered. This is done in a *design study* where each of the following steps are taken in turn:

1) Study the component drawing paying particular attention to the primary process (such as forging, die casting, etc.) and the tolerances on dimensions laid down by the designer.

2) Study the *planning sheet* in order to ascertain the manufacturing sequence of operations laid down by the planning engineer.

3) Study in detail the operation for which the fixture and tools are required, and decide by which process or processes the operation can be carried out. It will be necessary to estimate the *standard time* as described in Section 1.10.

4) Carry out a costing exercise in order to discover which of the technically feasible alternative processes is economically viable. Use the *break-even* analysis described in Section 1.3 for this purpose.

5) Complete the design study of the fixture by outlining the methods to be adopted for location, clamping, tool positioning, etc. This study

need only include sketched details of the design features showing their principle of operation.

The work done in this part of the design study is passed over to the tool design draughtsman who will complete the design on the drawing board and produce finished working drawings with each individual item detailed.

To illustrate this procedure, an example is now given of a design study for a milling fixture.

3.12 DESIGN STUDY FOR A MILLING FIXTURE

The design study is required for a fixture for the machining of the faces marked thus: ▽ on the component shown in Fig 3.20.

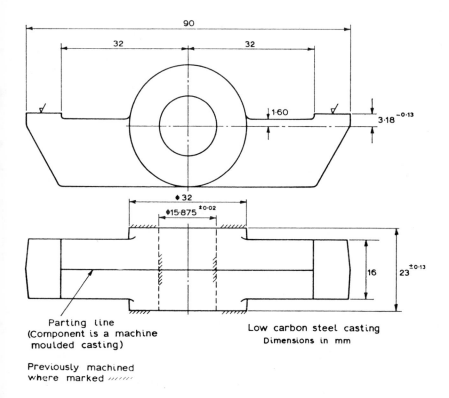

Fig 3.20 Component to be Milled.

The components are required in weekly batches of at least 3 000, the total lot size being substantial.

Design Study

1) The component is a casting with a mould parting line and some draft or taper which might warrant attention during location and clamping.
2) The sequence of operations is such that the bore and each end face of the 32 mm diameter boss have already been machined. This is to be followed by the machining operation on the pads.
3) Assuming a 40 hour week, the minimum production rate is
$\dfrac{3\,000}{40} = 75$/hour.

The pad faces could be machined by horizontal or vertical milling, or surface broaching. Assume that a horizontal milling machine and a surface broaching machine are available. It will now be necessary to check if these machines can achieve a production rate of 75/hour.

Milling

Assume a 100 mm diameter cutter having 12 teeth is used and the depth of the cut is 3 mm.

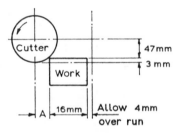

Fig 3.21   Milling Component.

From Section 1.7.
$$A = \sqrt{d(D-d)} = \sqrt{3 \times 97} = \sqrt{291} = 14 \cdot 15 \text{ mm}$$
$$\text{say } 15 \text{ mm}$$
$$L = l + A + \text{over-run} = 16 + 15 + 4 = 35 \text{ mm}$$

Cutter speed $= \dfrac{305}{100\pi} = 0 \cdot 97$ rev/s: $0 \cdot 90$ rev/s is available on the machine

At a feed of $0 \cdot 08$ mm/tooth, feed/rev $= 0 \cdot 08 \times 12 = 0 \cdot 96$ mm

$$\therefore f = 0 \cdot 96 \times 0 \cdot 90 = 0 \cdot 864 \text{ mm/s: } 0 \cdot 80 \text{ mm/s is available on}$$
$$\text{the machine}$$

$$\therefore \text{ Time } T = \dfrac{35}{0 \cdot 80} = 44 \text{ seconds.}$$

By synthesis the unloadings, loading time, etc., is found and added to the CR allowance, totalling 34 seconds.

∴ Standard time $= 44s + 34s = 78s = 1\cdot30$ min

$$\text{Output} = \frac{60}{1\cdot30} = 46 \text{ components/hour.}$$

This is below the required minimum rate but it is found that the machine a ble will accommodate two components as shown in Fig 3.22. Therefore h e fixture could be designed to hold two components.

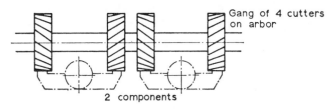

Fig 3.22

The allowances to the machining time are now found to be increased from 34s to 49s

∴ Standard time $= 44s + 49s = 93s = 1\cdot55$ min

$$\text{Output} = \frac{60}{1\cdot55} \times 2 = 77/\text{hour.}$$

This is satisfactory.

Broaching
Assume a cut/tooth of $0\cdot025$ mm and a cutting speed of $0\cdot05$ m/s.

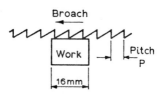

Fig 3.23 Broaching Component.

Pitch $P = 1\cdot77\sqrt{\text{length of cut}} = 1\cdot77\sqrt{16} = 1\cdot77 \times 4 = 7\cdot08$ say 7 mm

Number of teeth required $= \dfrac{3}{0\cdot025} = 120$ teeth

∴ Length of cutting portion required $= 120 \times 7 = 840$ mm

∴ Time $T = \dfrac{840 + 16}{50} = \dfrac{856}{50} = 17\cdot1$ seconds.

Using same allowance as before

Standard time $= 17\cdot1s + 34s = 51\cdot1s = 0\cdot85$ min

$$\text{Output} = \frac{60}{0\cdot85} = 71 \text{ components/hour.}$$

With a little increase in cutting speed this can easily be raised to 75 components/hour. This is satisfactory.

4) The milling fixture will be more expensive than the broaching fixture as it must be designed to hold two components. On the other hand the broach will be much more expensive than the milling cutters which could possibly be a stock item. An estimate must be made here of the various tooling costs involved (including the cost of the fixture) in order to make a reasonably break-even analysis. Past costing records which are accurate can be of enormous help here. Using the method outlined in 1.3:

	Milling process	Broaching process
a) Tooling cost (estimated)	£450	£500
b) Material cost/component	£0·72	£0·72
c) Operating labour cost	£1·44/h	£1·44/h
d) Standard time/component	0·78 min	0·85 min
e) Setting up labour cost	£2·88/h	£2·88/h
f) Setting up time	2h	4h
g) Machine overheads	300% of (c)	500% of (c)

Milling Machine

$$\text{Overheads} = \frac{300}{100} \times 1·44 = £4·32/h$$

$$\text{Fixed cost} = 450 + 2\,(2·88 + 4·32)$$
$$= 450 + 14·40 = £464·40$$

$$\text{Variable cost/component} = \left(1·44 \times \frac{0·78}{60}\right) + 0·72 + \left(4·32 \times \frac{0·78}{60}\right)$$

$$= \left(5·76 \times \frac{0·78}{60}\right) + 0·72 = £0·7949$$

Variable cost/1 000 components $= £794·90$

Broaching Machine

$$\text{Overheads} = \frac{500}{100} \times 1·44 = £7·20/h$$

$$\text{Fixed cost} = 500 + 4(2·88 + 7·20)$$
$$= 500 + 40·32 = £540·32$$

$$\text{Variable cost/component} = \left(1·44 \times \frac{0·85}{60}\right) + 0·72 + \left(7·20 \times \frac{0·85}{60}\right)$$

$$= \left(8·64 \times \frac{0·85}{60}\right) + 0·72 = £0·8424$$

Variable cost/1 000 components $= £842·40$

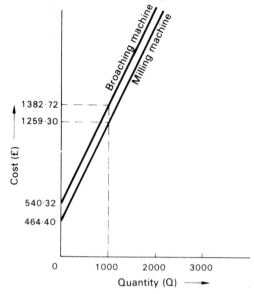

Fig 3.24 Break-even Chart.

In this instance both the fixed and variable costs of the broaching process are a little higher than for the milling process, therefore it is obvious that at the required production rate the milling process should be used. However, the break-even analysis is shown as an exercise. Had the components been required as fast as possible the situation would be quite different. Using more expensive tooling with a duplex fixture, the broaching process would be much faster than milling. Then a break-even analysis might well have favoured broaching due to the decrease in variable costs although fixed costs would have been higher.

5) Having decided upon the milling process, the design features of the fixture can now be studied. These are (a) location, (b) clamping, (c) tool setting, (d) machine capacity and operator safety, (e) outline of construction of fixture.

(a) *Location.* The component can first be located on a pin of $15.854^{-0.05}$ mm diameter which allows any component bore within tolerance to locate easily and also allows adequate control over the $3.18^{-0.13}$ mm dimension. This location removes four degrees of freedom.

The fifth degree of freedom can be removed using two pegs as shown in Fig 3.25, which locate upon unmachined cast faces.

These pegs will need to be floating and interconnected to allow for any mal-alignment of the cast faces. Location (and hence clamping) is not envisaged upon machined boss face A (Fig 3.25), as the support and clamping is required as close to the machining zone as possible for rigidity.

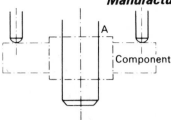

Fig 3.25   Locating Component.

The sixth degree of freedom can be removed by location pegs underneath the component as shown in Fig 3.26.

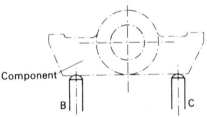

Fig 3.26   Locating Component.

These locaters would contact the casting approximately on the mould parting line, and clearance may have to be provided on the pegs to give a solid seating. Any variation in casting size and shape must be accommodated. This could be done by making peg C fixed and peg B adjustable, this being designed so that the component is always pressed against peg C, either by automatic means or by the operator. The casting will then always be approximately symmetrical about horizontal centre line.

All location surfaces should be hardened. Burrs from the milling operation must not prevent withdrawal of the component from the fixture. The component should be easy to unload provided peg B is easily retractable. Foolproofing pins will be needed, as shown in Fig 3.27, to prevent the component being loaded on to the main location pin upside down.

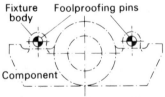

Fig 3.27   Foolproofing Pins.

These pins can be accommodated in the main structure of the fixture.
(b) *Clamping*. The action of the clamps should be such as to thrust the component back against the location pegs which, being floating, would

adjust themselves to suit any discrepancy in the casting shape. This arrangement should give maximum rigidity. Any clamps used should clear the cutters.

As two components are being accommodated in the one fixture, it would be desirable if the clamps could be actuated from a single control. Furthermore the clamps should preferably be floating to accommodate any discrepancy in casting size because they will be bearing upon a rough cast face. This can be done in the way shown in Fig 3.28.

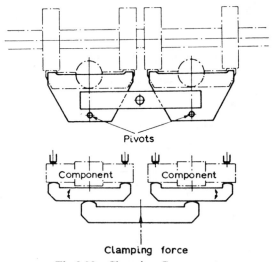

Fig 3.28 Clamping Components.

The clamping force could be imparted by means of a screw or cam actuated by the operator, or could be pneumatically controlled. The final designed arrangement will depend upon the cost budget which was used for the costing analysis. At this stage only design principles are being committed to paper.

In all clamping designs one should be careful that previously machined faces are not being marked. This is not a consideration here.

(c) *Tool Setting.* A hardened and ground setting block is required, to be used in conjunction with feeler gauges (as described in Section 3.6(c)). In this case only one cutter needs setting in position relative to the fixture, then the rest of the gang of cutters are positioned by setting collars on the machine arbor. This is shown in Fig 3.29.

Hardened and ground tenon blocks will be required on the base of the fixture (as described in Section 3.6(d)). These will make for quick and easy setting on the machine table although very accurate alignment of the machine and fixture axes is not necessary in this case.

(d) *Machine capacity and operator safety.* The total estimated width of the fixture should now be checked against the Machine Capacity

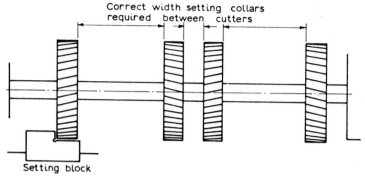

Fig 3.29   Setting Block for Milling Cutter.

Chart, and must not be greater than the maximum distance from the face of the machine column to the inside of the arbor bearing support, as shown in Fig 3.30.

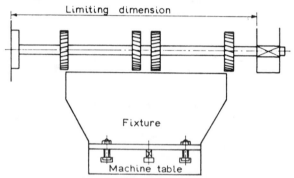

Fig 3.30   Fixture position upon Machine.

The fixture may be designed wider than the machine table if desired using the construction shown. The total required traverse should also be checked against the stroke of the machine table. In this case with a traverse of only 35 mm, there would be a more than adequate stroke. Also tenon slot dimensions on the table should be noted when designing fixture bolt slots.

Accurate *machine capacity charts* showing every necessary dimension and detail of each machine used for production purposes should be filed in the Jig and Tool Drawing Office. An example of such a chart is shown in Fig 3.31.

When considering operator safety a difficult problem usually arises with milling fixtures. This is to do with the siting of the clamp operating point. If a manually operated screw clamping arrangement is to be used as seems likely in this case, then the clamp operating point should preferably be on the opposite side of the fixture to the cutter as shown in Fig 3.32.

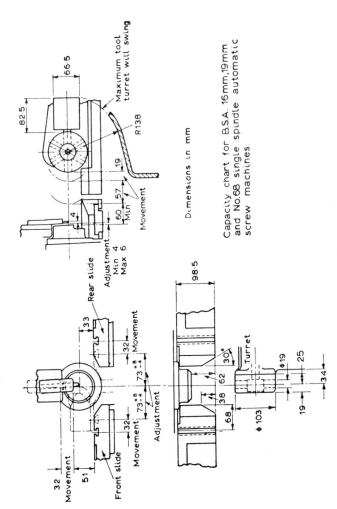

Capacity chart for B.S.A. 16mm, 19mm and No.68 single spindle automatic screw machines.

Dimensions in mm

Fig 3.31 Capacity Chart for B.S.A. Single Spindle Automatic.

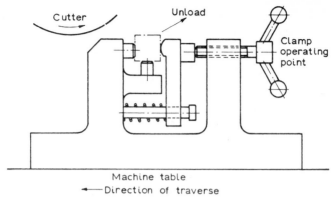

Fig 3.32   Clamp operating point relative to Cutter.

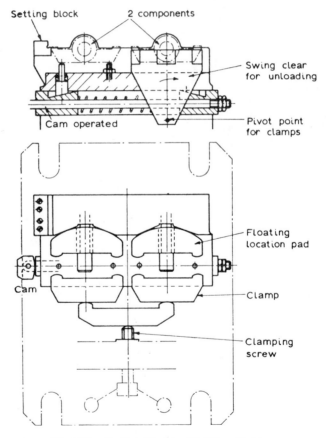

Fig 3.33   Outline Sketch of Milling Fixture.

The arrangement shown means that the clamps are taking the cutter thrust which is wrong in principle. However, with rigid and robust clamps this is not of vital importance. Remote control devices for operating clamps can be arranged if absolutely necessary; these are manipulated on the opposite side of the fixture to the cutter and the clamps. They will add considerably to the cost as they are quite elaborate in construction. Cutters must be guarded, and preferably a limit switch should be arranged to stop the cutters rotating when the table traverse is completed.

(e) *Outline of Construction of Fixture.* The main body of the fixture will best be manufactured from cast iron in the form of a casting for the reasons outlined in Section 3.6(f).

All the design ideas described up to this stage can now be put together and a picture of the milling fixture begins to emerge. This composite picture can be sketched to complete the design study and might look like Fig 3.33.

It will be noted that the location pins which removed the fifth degree of freedom have been incorporated into one unit which is floating and allows for variation in the size of the casting. There is now sufficient

	DESIGN STUDY SHEET	Form No.
COMPONENT	DRG. No.	SKETCHES
MATERIAL	PRODUCTION QUANTITIES	
OPERATION	MACHINE	
LOCATION		
CLAMPING		
LOADING METHOD		
SECURING		
CONSTRUCTION		
SAFETY		
CUTTING TOOLS		

Fig 3.34 Design Study Sheet.

information in this design study to enable a jig and tool designer to complete the design on the drawing board and issue working drawings to the Toolroom where the fixture will be manufactured.

Standard forms could conveniently be used for design studies, a suggested one being shown in Fig 3.34, calculations and sketches accompanying the form.

Exercises 3

1. Complete the general assembly drawing of the milling fixture evolved from the design study in Section 3.12.

2. What are the principal design points to observe for the location of a component in a fixture. Give an example of each.

3. What are the principal design points to observe for the clamping of a component in a fixture. Give an example of each.

4. The component in Fig 3.14 is to have the 13 mm slot and 9 mm thick forks milled before the turning operation shown. Design a string milling fixture for this purpose to take six components when fully loaded. State the type and size of milling cutters used.

5. A standard milling machine vice fitted with special jaws can be used very effectively as a milling fixture. Show an example of this.

6. A component 20 mm thick × 100 mm diameter is to have six equispaced 5 mm diameter holes drilled on an 80 mm PCD. Sketch the outline of a design of an indexing drilling fixture suitable for this drilling operation to be carried out on a single spindle drilling machine.

7. The component shown in Fig 3.35 is to be cylindrical ground on the outside to the diameter given, and has already been ground in the bore

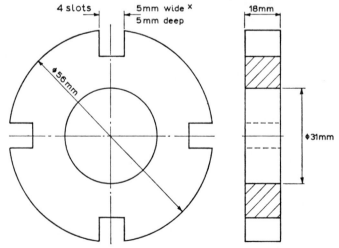

Fig 3.35

and on the faces. Design a grinding mandrel which is suitable for this operation to accommodate eight components.

8. Design a manually operated clamp for use where projections on the work make the use of a simple, direct operation clamp impossible. Assume the outline of the component is as shown in Fig 3.36.

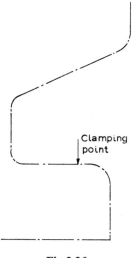

Clamping point

Fig 3.36

9. A simple, cast iron bell crank lever having 150 mm long × 15 mm thick arms at 90° to each other is to have a 25 mm diameter hole bored in its 50 mm diameter × 25 mm thick boss on a capstan lathe. Assume the boss has been machined previously on each face. Design a turning fixture for this operation which will hold one component.

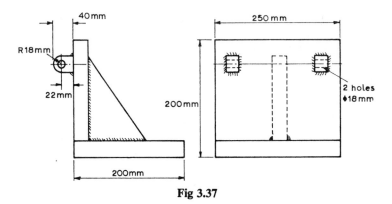

Fig 3.37

10. The bracket shown in Fig 3.37 is fabricated by electric arc welding from five pieces of 20 mm thick mild steel. Design a welding fixture for locating and clamping the individual pieces.

11. The completed bracket shown in Fig 3.37 has been machined on the base and requires the two holes of 18 mm diameter to be machined. These holes must lie on the same axis which must be parallel to the base.

Design a fixture for this operation assuming the brackets are required in occasional batches of 100.

Further Reading

1) Lissaman A. J. and Martin S. J. *Principles of Engineering Production.* Hodder and Stoughton.

2) Jones E. J. H. *Jig and Tool Design.* George Newnes Ltd.

3) Colvin F. H. and Hass L. L. *Jigs and Fixtures.* McGraw-Hill Book Company Inc.

4) Town H. C. *Cutting Tools, Jigs and Fixtures.* Odhams Press Ltd.

5) Kempster M. H. A. *Introduction to Jig and Tool Design.* Hodder and Stoughton Ltd.

6) Parsons S. A. J. *Production Tool Equipment.* Cleaver Hulme Press Ltd.

7) Kempster M. H. A. *Principles of Jig and Tool Design.* Hodder and Stoughton Ltd.

CHAPTER 4

Machine Tools

4.1 MACHINE TOOLS USED FOR QUANTITY PRODUCTION

THE MORE complex machine tool systems used in quantity production will be considered in this chapter. As labour cost is often the greatest element of cost in the total cost of a product, these machines are mostly automatic, or semi-automatic where the machine operator is virtually a material feeder, in order to reduce the labour cost to the minimum. It is assumed that the reader has a background knowledge and experience of standard machine tools such as centre lathes, milling machines, shaping machines, drilling machines and cylindrical and surface grinding machines, which in their standard form are not ideal production machines.

Several production type machine tools which require some degree of operating skill (if sometimes only a small amount) such as capstan lathes, turret lathes (horizontal and vertical), and boring mills for example, will not be considered here. These machines are suitable for short runs or long runs, and often the operator has sufficient skill to set up the machine before commencing production. The operators of such machines can in no way be classed as material feeders. An understanding of the principles used in tooling up the machines mentioned above, and of the various tooling devices available (such as roller box tools for example), will be of great help in understanding the principle of operation of some of the automatic lathes. Sections 4.12, 4.13 and 4.14 are concerned with copying systems as applied to production work, although to some extent this area of work has been overtaken by numerically controlled machining systems (see Chapter 6).

4.2 SEMI-AUTOMATIC, MULTI-TOOL CENTRE LATHES

On this type of lathe the work, if solid, is mounted between centres, or if bored, is mounted upon a mandrel. Several tools are used so that the turning is completed in one cycle. The traversing of the tools across the work-piece is carried out automatically, and the operator merely loads and unloads the work between the centres. Therefore, depending upon the cycle time, an operator might feed more than one machine.

The turning tools are carried in a series of tool boxes at the front and back of the machine, the tools being carried on cam operated slides which traverse on dovetail or round slideways. The rear tools are usually arranged for cross traversing and carry out the plunge turning operations such as undercutting, chamfering or forming. The front tools then traverse longitudinally and carry out the plain turning of diameters. One feed can be chosen for turning and one for fast return.

The general configuration of the tools is as shown in Fig 4.1.

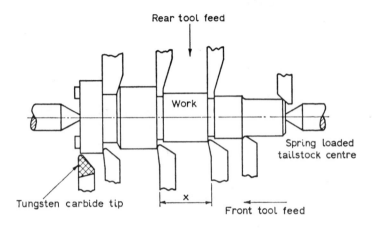

Multi-tool centre lathe tool set up

Fig 4.1   Multi-tool Centre Lathe Tool Set-up.

The total machining time per component is the time to turn the longest shoulder length, this in the example shown being length x. If uneven wear takes place on the tools then inaccuracies will quickly develop in the work. This can be overcome to some extent by using high speed steel tools for the smaller diameters, and tungsten carbide tools for the larger diameters. This will also allow higher cutting speeds to be used hence increasing the production rate.

The work may be castings, forgings or bar which has already been centred and machined to length by facing. A spring loaded tailstock is used for quick release and loading of the workpiece. Consideration must be given to the method of driving the component between centres from the headstock. This must be fast, simple and positive. Therefore the method used upon centre lathes of fastening a driving carrier upon the workpiece is not suitable. Advantage must be taken of any holes, slots or lugs (as in the component shown in Fig 4.1) in the workpiece to obtain a positive drive from the headstock driving plate. In such a case the sequence of operations would be arranged so that the driving medium is machined prior to the multi-turning operation. Also a special driving plate would be required to be designed in the jig and tool drawing office for each batch of components to be turned. In the case of bar work the component can be driven from the outside diameter using a quick release type of chuck which simply drives the bar between centres from two jaws. Alternatively, if the component is to be turned over its full length in one operation a commercial driver of the 'Kosta' type can be used. This has driving pins which press against the end face of the bar, and it may be necessary to fit a pressure gauge to the tailstock in order to ensure that the pressure between the driving pins and the tailstock centre is not excessive. Long, slender work will need steadying during machining, and the use of these will increase the set up time and hence increase the fixed costs.

Consideration must be given to the power required for a multi-turning operation. This in many cases will be very high and it would be wise to check that the motor is not being overloaded.

Power required for a single tool
$$= W = PcwV$$
where P = Specific cutting pressure (N/mm^2)
$\quad c$ = Tool feed (mm/rev)
$\quad w$ = Depth of cut (mm)
$\quad V$ = Cutting speed (m/s).
For several tools in a multi-tool set up
$$\text{Total power} = \Sigma(Pcw) \times V$$

Example 4.1

A 102 mm diameter mild steel bar is being turned at 1·70 rev/s on a multi-tool centre lathe having nine tools. For all tools the depth of cut is 5 mm, the feed is 0·255 mm/rev, and the specific cutting pressure is 1 545N/mm^2. Calculate the total power consumed for the operation.

Solution

Cutting speed $V = \text{rev/s} \times \pi d = 1·70 \times 3·142 \times \frac{102}{1000} = 0·545 \text{ m/s}$
Specific cutting pressure $P = 1\ 545\text{N/mm}^2$
$$\Sigma Pcw = 9 \times 1·545 \times 0·255 \times 5 = 17·730 \text{ kN}$$
$\therefore$ Total power $= \Sigma Pcw \times V = 17·730 \times 0·545 = 9·68 \text{ kW}$

4.3 AUTO LATHES

This type of production lathe is sometimes called an automatic turret lathe or chucking lathe. It is similar in configuration to a capstan or turret lathe and will take work such as bar billets, forgings or castings which are suitable for a hydraulically or pneumatically operated chuck. An operator is required only to load and unload the work as the tools are automatically presented to the work on cam operated cross slides and turret. Up to four machines can be manned by one tool setter and one operator, and on the average, savings in labour costs are to the order of 40% to 50%. Setting costs are similar to those on a capstan lathe as the tooling used is very similar; setting of the cams however must also be carried out for each new batch of components.

Consistent, high quality work up to 0·60 m diameter can be produced on either short or long runs, and the operator suffers much less from fatigue than a capstan lathe operator. The general layout of an auto lathe is as shown in Fig 4.2.

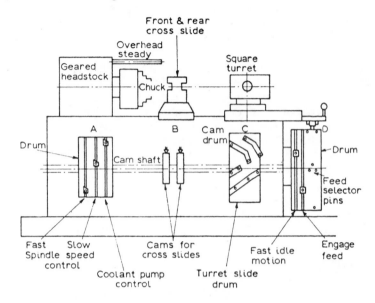

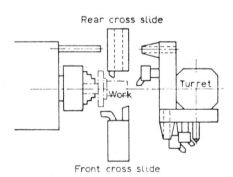

Fig 4.2   Diagram of Auto Lathe.

This diagram shows simply the general relationship of the slides and tools to each other and is not intended to represent an authentic layout. The machine has two independent cross slides which are separately operated from adjustable cams shown at *B*. The turret slide is operated by the adjustable cam drum C, and the turret indexes to the next face at the end of each stroke. Each cross slide can be traversed when a particular set of turret tools is traversing, i.e., overlapping of operations.

The cams and drums at the front of the machine can be adjusted at the setting up stage by removing the guarding covers. One fast and one slow spindle speed is available for any component, these being selected by pick-off gears. The dogs on drum A are adjusted to change the speed as required during the machining cycle. The third dog on A switches the coolant supply on and off.

Selector pins can be arranged in the appropriate holes on drum D to allow the correct feed for each machining operation to be automatically selected. Dogs on the drum also control the times at which the turret traversing feed engages for forward traversing and fast idle return traverse.

It will be seen from the diagram that multi-tool set ups are used, the steady bar being used with multi-tool knee toolholders enabling high rates of metal removal to be attained where required.

Where very long production runs justify it, automatic loading magazines can be fitted thus making the system fully automatic. This combined with tungsten carbide, pre-set multi-tools holders and as much overlapping as possible, will give a high rate of production.

4.4 TURRET TYPE SINGLE SPINDLE AUTOMATICS

The automatic lathe is the logical development of the capstan lathe. The other main type of single spindle automatic in addition to the turret type is the sliding head type which will be considered in Section 4.5. Multi-spindle automatics will be considered in Section 4.6.

Turret type auto's are sometimes known as automatic screwing machines because of the facility with which small screwed parts can be produced upon them in high quantities. Some can accept bar up to 50 mm diameter in a collet chuck, and the general configuration is that of a capstan lathe except that a third overhead tool slide is provided. This is usually used for parting off if the other two independent cross slides are used for other tools. Also the other main difference is that a six station round tool turret is provided which indexes about a horizontal, instead of vertical, axis.

The general layout of a turret automatic is shown in Fig 4.3. A simple plan view showing the principle of operation of the main features is shown at Fig 4.4.

It is not the intention in this book to analyse machine tool design. However, some knowledge of the principle of operation of an automatic is required before one can understand how it can be cammed and tooled up for quantity production.

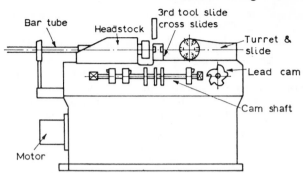

Fig 4.3   General Layout of Turret Automatic.

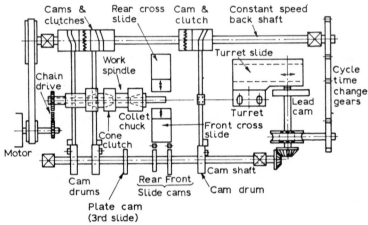

Fig 4.4   Plan view showing Principle of Operation of a Turret Automatic

Specially manufactured plate cams are used to operate the tool slides and turret slide. This latter cam is called the lead cam. They all originate as circular discs and are machined to the required profile. Each tool slide carries one tool only, and the turret carries six; only one turret station being in operation at once. Therefore the slide cams have a single lobe, and the lead cam will have as many lobes as there are turret stations being used. The radial throw (lift) of each cam lobe is equivalent to the required length of travel of the tool for which it is designed. The feed of each tool is also controlled by the rate of lift of the cam lobe. The turret indexes automatically one (or two if required) station, by means of a geneva plate mechanism, at the end of each reverse stroke of the turret slide.

Reference to Fig 4.4 shows the cam shaft at the front of the machine. At the back of the machine is the backshaft which rotates at a constant speed (usually 2 rev/s or 4 rev/s). The backshaft carries dog clutches in three positions which are operated through drum cams; these in turn being actuated through levers from cam drums carried on the front cam shaft. By setting trip dogs in the correct angular positions on the cam drums, the dog clutches on the back shaft can be made to operate when required in the cycle of operations. These three clutches when operated will cause (a) the turret to index one (or two) stations, (b) the collet chuck to open, hence the bar feeds forward to the bar stop in the turret, and the collet closes again gripping the bar, and (c) the spindle speed to change from the selected fast speed to the selected slow speed, or vice versa, respectively. The backshaft rotates one revolution while any one of these idle operations is being carried out.

The cam shaft is driven from the backshaft through cycle time change gears. (See Fig 4.4.) This is usually a compound train of pick off spur gears which is changed to suit each new component being tooled up. One revolution of the front cam shaft produces one component. As the backshaft rotates at a constant speed therefore the speed of rotation of the cam shaft can be varied to suit each new component by means of the cycle time gears. Hence the cycle time can be varied. A limited number of cycle time change gears are provided, and reference to the makers' handbook will show what cycle times are available at any of the spindle speeds available on the machine.

Many clever, special devices are available to increase productivity, such as slot sawing attachments, cross drilling attachments, etc. These each require a cam to operate them, and provision is made on the front cam shaft for this purpose.

Turret Automatic Cam Design

In order to design cams for a turret automatic, it is best to work to a set procedure consisting of several steps. We will work through one example in detail in order to show the principle. The examples referred to in this section are provided by courtesy of *B.S.A. Tools Ltd., Birmingham, England.*

Example 4.2

The component shown in Fig 4.5 is required in large quantities and is to be machined on a turret automatic. Design and draw a set of plate cams for this purpose. Show the tool layout.

Procedure. (When following these steps refer to Fig 4.6 which shows the cam design sheet.)

1) *Determine rev/s of Workspindle.*

First select suitable cutting speed, say 2·50 m/s for 9·50 mm diameter brass using high speed steel tools.

$$\therefore \text{rev/s} = \frac{\text{cutting speed}}{\pi d} = \frac{2\cdot50 \times 1\,000}{3\cdot142 \times 9\cdot50} = 84 \quad \begin{array}{l}\text{Select nearest suitable}\\\text{spindle speed on mach-}\\\text{ine fast range.}\end{array}$$

$\therefore$ In this case 84 rev/s is available.

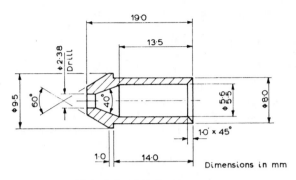

Material:- ⌀ 9·5 Brass
Fig 4.5

(Note. One other speed is available on the slow range if required, for screw cutting for example.)

2) *Determine Sequence of Operations.*

Overlap operations (i.e., carry out more than one operation at once) where possible to save time. In Operations 4, 5 and 6 overlap forming operation from front cross slide, with drilling, coning and reaming from turret.

3) *Determine Throw (Travel) of each Tool.*

Add up to 0·40 mm approach to avoid damage to tools.

4) *Determine Feed/Rev of Tool.*

The choice of feed will depend upon the type of material, the operation being carried out and the type of tool used. Experience is required in order to determine the best value.

5) *Determine No. of Revolutions required for each:*

a) *Machining Operation.* Work spindle revs $= \dfrac{\text{Throw of tool (mm)}}{\text{Feed of tool (mm/rev)}}$

e.g., Operation No. 2 Revs $= \dfrac{3\cdot00}{0\cdot15} = 20$

b) *Idle Operation.* Enter in overlapped column where appropriate. (Note. Backshaft rotates at constant speed of 4 rev/s.)

One revolution of backshaft required to (i) Index turret; (ii) Feed stock; (iii) Change from fast to slow speed or vice versa.

Thus to index, feed stock or change speed takes $\frac{1}{4}$ second.

This will give $84 \times \frac{1}{4} = 21$ workspindle revs (add five extra revolutions say, to facilitate setting of trip dogs, where these can be added after correcting total number of revs.)

c) *Drop Back to Clear.* When the turret is indexed, the amount of straight withdrawal before the turret starts to revolve is approx. 19 mm. ∴ In Operation 4 for example the 2·38 mm diameter drill must be withdrawn 3 mm before the turret can be indexed. Similarly with the parting off tool in Operation 7 before feeding out stock for next component.

6) *Determine Cycle Time* (i.e., time to produce one component).
Disregarding overlapped operations, add up the total of revolutions
required for each operation. This equals 532. Check in machine handbook under 84 rev/s spindle speed, the nearest revs/piece obtainable
using standard cycle time change gears. In this case it is 583 revolutions.

This gives cycle time $= \dfrac{583}{84} = $ approx. 7 seconds.

Turret Automatic — All cams have 1:1 ratio with slides or turret e.g. 25mm rise (or throw) on cam moves tool slide 25mm

Cam design work sheet. No. 48 SS Automatic screw m/c. Short stroke

Work spindle speeds rev/s		Cutting speeds m/s	
Fast	RH 84	Turning	2.5
	RH		
	RH		
Slow			

Dimensions in mm. Production time - 7 sec. Matl:- Brass ⌀9.5mm

O.N.O.	Distance between turret and chuck (Minimum 46.0mm / Travel 35.0mm)	Sequence of operations	Throw in mm	Feed mm/rev	Revs of work spindle — For each opern	Over-lapped opern	Corrected	100's
	17.5 Cut mm down	Feed stock	Dwell		21		26	4½
1	Workpiece	Index turret (one station)			21		26	4½
2	16.0 mm	Centre end of bar	3.0	0.15	20		29	5
		Dwell to clean up			8		6	1
		Index turret (one station)			21		26	4½
3	6.0 mm	Turn ⌀8.0mm & Drill ⌀5.3mm	14.5	0.13	114		117	20
		Dwell to clean up			8		6	1
		Index turret (one station)			21		26	4½
4	9.5 mm	Drill ⌀2.38mm hole	7.1	0.08	89		92	16
		Drop back to clear			6		9	1½
		Form (overlapped)	2.5	0.013		200	(204)	(35)
		Dwell to clean up			6			(1)
		Index turret (one station)			21		26	4½
5	9.5 mm	Cone bottom of ⌀5.3mm hole	3.4	0.07	45		50	8½
		Dwell to clean up			8		6	1
		Index turret (one station)			21		26	4½
6	9.5 mm	Ream ⌀5.5mm & cone	13.0	0.25	52		53	9
		Dwell to clean up			8		6	1
		Index turret (overlapped)				26		(4½)
7		Cut off rear cross slide	1.5	0.05	30		32	5½
		Cut off rear cross slide	0.5	0.08	6		9	1½
		Clear cut-off tool			12		12	2
		Totals			532		583	100

Fig 4.6 Cam Design Work Sheet for Turret Automatic.

7) *Correct Revolutions/Operation.*

Additional revolutions given as result of (6) above, i.e., $583 - 532 = 51$ revolutions. These should be allotted to those operations which will benefit most, e.g., increase 'indexing turret' from 21 to 26, or increase 'centre end of bar' from 20 to 29, etc. Total corrected revolutions should equal 583.

8) *Convert Revolutions/Operation into Hundredths.*

583 revs $= \dfrac{100\text{'s}}{100}$ of cam periphery. Therefore multiply each number of revolutions in corrected revs column by $\dfrac{100}{583}$ to give 100's, e.g., 26 revs

$= \dfrac{100}{583} \times 26 = 4\frac{1}{2}$. (To nearest $\frac{1}{2}$ hundredth.)

Total of hundredths column (ignoring overlapped operations) should, of course equal 100.

9) *Draw the Cams.*

In practice the lobes can only be drawn when '*cut down*' (i.e., reduction to maximum radius of lead cam lobe) has been determined for each operation. Tool dimensions must be known to do this. High point of the lobe for each cross slide cam coincides with circumference of blank.

The final tool layout is shown at Fig 4.7 and the three cams are drawn at Fig 4.8. Note that the third slide is not used in this case as the parting off can be done from the rear tool slide.

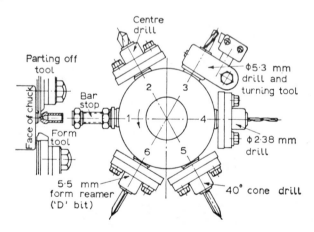

Fig 4.7   Tool Layout for Turret Automatic.

The cams are drawn superimposed, one on top of the other in order to show the working position to which they will be set on the camshaft.

This type of component could be machined consistently to the tolerance required over a long period of time. One setter and one operator can keep several auto's running, the operator's task being simply that of feeding the machines with bar stock. Hence, variable costs are comparatively low, but fixed costs will be higher than for a capstan lathe say, due to the cost of cams and the longer time required for setting up. These machines are ideal for short length work having a limiting diameter to length ratio of 1 to 3.

A second example of cam design is given where the third slide is required for parting off.

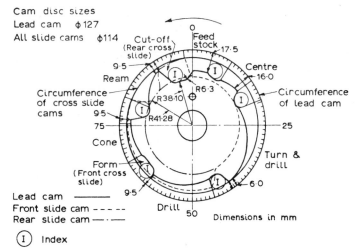

Fig 4.8 Cam Layout for Turret Automatic.

Example 4.3
The component shown in Fig 4.9 is to be machined upon a turret automatic. Design and draw suitable cams. Show the tool layout.

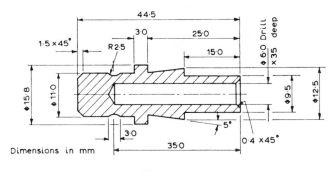

Material :- ⌀ 15·8 Brass

Fig 4.9

Solution

The cam design sheet is shown at Fig 4.10. Note that a cutting speed of
2·50 m/s is used at 15·80 mm diameter giving the nearest available

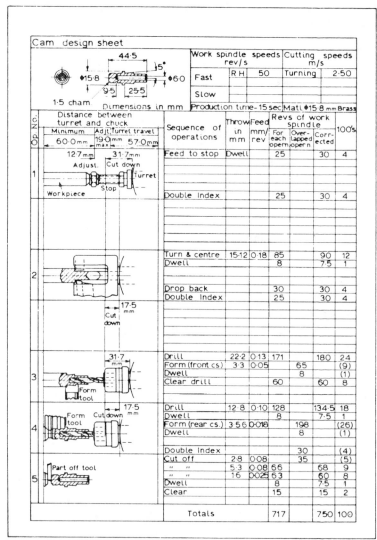

Fig 4.10   Cam Design Work Sheet for Turret Automatic.

spindle speed of 50 rev/s. The drilling operation is carried out in two
parts (Operations 3 and 4) in order to clear the drill flutes of brass
swarf. This is good practice where the drill hole is long in relation to its

diameter. The parting off operation (Operation 5) is broken down into three parts because (a) part of it is overlapped with the drilling and forming operation and does not count in the final cycle time, and (b) the feed rate is changed as the tool approaches within 1·60 mm of the work axis. Double indexing is used as three tools only are required in the turret.

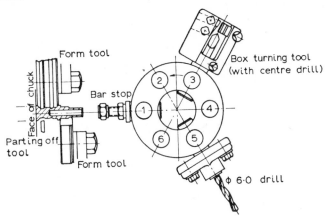

Fig 4.11 Tool Layout for Turret Automatic.

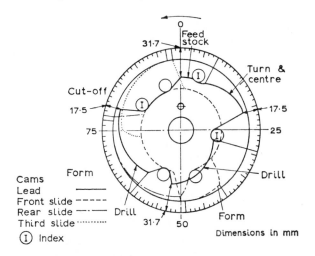

Fig 4.12 Cam Layout for Turret Automatic.

The backshaft speed is 2 rev/s. Hence each idle operation takes $\frac{1}{2}$ second. At a spindle speed of 50 rev/s, 25 revolutions of the spindle will elapse during that time. The number of revolutions required per component totals 717. From the handbook the nearest number of revolutions available using standard cycle time change gears at 50 rev/s

is 750. 750 revolutions then $= \dfrac{100}{100\text{'s}}$ of cam periphery giving a cycle

time of 15 seconds. The revolutions for each appropriate operation are altered accordingly.

The final tool layout is shown at Fig 4.11 and the four cams are drawn at Fig 4.12.

It will be appreciated that the procedure given is a simplified version of that used in practice, and from that given in the manufacturers' handbook. This is done in order to make for an easier understanding of the principles involved.

4.5 SLIDING HEAD TYPE SINGLE SPINDLE AUTOMATICS

This type of machine is entirely different in conception than the turret type, having an entirely different machining principle than any yet considered. It was developed in Switzerland for manufacturing quantities of small, precision components, and it is often referred to as a Swiss Auto. The machine accepts bar stock in a collet chuck, and has four major features:

1) A sliding headstock through which the bar passes and in which it is held by a collet.

2) A tool bracket which supports four or five tool slides, and also a guide or steady bush for the bar stock.

3) A front camshaft holding the cams which control the tool slides and headstock movements.

4) A feed base (instead of turret slide) which may be multi or single spindle. This is used for 'hole operations' such as drilling, screwing etc.

The general layout of a sliding head automatic is shown in Fig 4.13. A simple plan view showing the principle of operation of the main features is shown at Fig 4.14. A front view of the arrangement of the tool slides is shown at Fig 4.15.

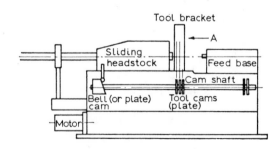

Fig 4.13   General Layout of Sliding Head Automatic.

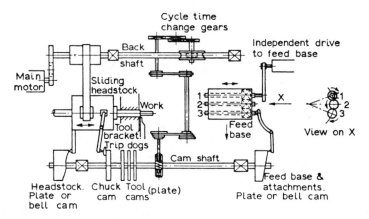

Fig 4.14 Plan view showing Principle of Operation of a Sliding Head Automatic.

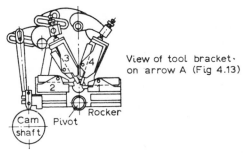

View of tool bracket·
on arrow A (Fig 4.13)

Fig 4.15 Arrangement of Tool Slides of a Sliding Head Automatic.

Again, specially manufactured cams are used to operate the various elements of the machine. The headstock moves the rotating bar past tools which are fixed longitudinally, but which can move independently in and out transversely. Hence, if the tool is moved in by a cam lobe and then held stationary by a concentric portion on the cam, a plain, parallel diameter will be turned on the workpiece as the bar is fed longitudinally past the tool. The headstock is moved in a 1 – 1 ratio by means of a bell (drum) cam, or a plate cam through a bell crank lever. The tool slides are operated through adjustable levers from the plate cams, giving a cam rise to tool movement ratio other than 1 – 1, say 1 – 1, 2 – 1 and 3 – 1 for example.

Reference to Fig 4.13 shows that tools number 1 and 2 are mounted on the front and rear of a rocker operated by a single cam. This means that these two tools cannot be used together. Tool number 1 is recommended for plain turning only, the other tools being used for parting off, forming etc.

The tool bracket holds a guide bush through which the bar passes giving a high degree of stability and steadiness during machining. Therefore work having up to a 1 – 10 diameter/length ratio can be successfully turned.

The cam shaft is driven through cycle time change gears from the constant speed back shaft as shown in Fig 4.14. One revolution of the cam shaft produces one component, hence various standard component cycle times are available by changing these gears, as with the turret automatic.

Drilling, tapping, etc., can be carried out using a cam fed, independently driven feed base. Hence, use can be made of the *differential principle* of machining, i.e., while the work revolves at one chosen cutting speed, the cutting speed for screwing say can be lowered by rotating the tap or die head on the feed base at a different speed. The difference between the work speed and the feed base spindle speed is the cutting speed for the operation. When cutting RH threads the die head revolves faster than the work spindle; for LH threads it revolves slower than the work. By the same token, drilling cutting speeds can be increased without a change in the work spindle speed, by rotating the drill in the opposite direction to the work spindle. Hence the cutting speed is the sum of the two spindle speeds. On multi-spindle feed bases, each spindle can be indexed in line with the work axis when required.

No special bar stop is required. The parting off tool is left in the forward position after cutting through the work. The chuck opens and the headstock feeds back over the bar to re-grip it as the chuck closes, while the bar remains pressed against the parting off tool. The tool then retracts and the next work cycle commences. Trip dogs are set in the correct angular position on the cam drum to operate the chuck when required.

Again as with the turret auto, special attachments are available to increase the productivity of the machine.

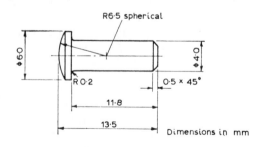

Material :- ⌀ 6·34 Brass

Fig 4.16

Sliding Head Automatic Cam Design

In order to illustrate the design procedure we will work through an example. The example used in simplified form is provided by courtesy of *A. C. Wickman Ltd., Coventry, England.*

Example 4.4

The component shown in Fig 4.16 is required in large quantities, and is to be machined on a sliding head automatic. Design and draw a set of plate cams for this purpose. Show the tool layout.

Procedure

(When following these steps refer to Fig 4.17 which shows the cam design sheet.)

1) *Determine rev/s of Workspindle.*

As for turret auto.

$$\text{rev/s} = \frac{2 \cdot 50 \times 1\ 000}{3 \cdot 142 \times 6 \cdot 34} = 125$$

∴ Select 120 rev/s giving 2·40 m/s cutting speed at 6·34 mm diameter

2) *Determine Sequence of Operations.*

Overlap operations where possible, e.g., Operations 5, 17 and 18.

3) *Determine Throw of each Tool.*

Allow 0·5 to 1·0 mm approach to avoid damage to tools. Throw of tool must be multiplied by cam ratio to give cam rise or fall.

4) *Determine Feed/Rev of Tool.*

As for turret auto.

5) *Determine No. of Revolutions Required*

for (a) each machining operation and (b) each entry or exit of tool. Enter in overlapped column where appropriate.

$$\text{e.g., Operation No. 4. Revs} = \frac{4 \cdot 00}{0 \cdot 64} = 6 \cdot 2$$

6) *Determine No. of Degrees for each Idle Operation*

Use allowances given in Fig 4.17.

7) *Determine Total of* (6) *above*

In this case the total = 88°. Therefore this leaves 360° − 88° = 272° as total of production operation degrees.

8) *Determine No. of Degrees for each Productive Operation*

Using total of production revolutions, i.e., 283, this is done by proportion. e.g., Operation No. 7, number of degrees $= \dfrac{118}{283} \times 272° = 113°$.

9) *Determine Cycle Time*

Total number of revolutions to produce one piece (in 360° of cam movement)

$$= \frac{\text{Production Revolutions}}{\text{Production Degrees}} \times 360$$

$$= \frac{283}{272} \times 360 = 374 \text{ revolutions}$$

Check in machine handbook under 120 rev/s spindle speed the nearest revs/piece obtainable using standard cycle time change gears. In this case 374 revolutions are available. If not, any additional revolutions must be allocated and the corrected revolutions column completed as described under (7) in the procedure for designing cams for turret auto's. The degree column will remain unaltered. The cycle time is calculated

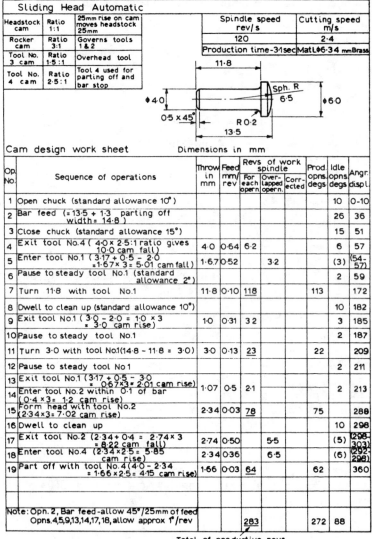

Fig 4.17   Cam Design Work Sheet for Sliding Head Automatic.

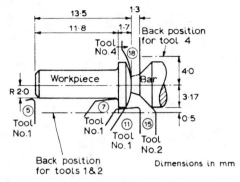

Fig 4.18 Tool Layout for Sliding Head Automatic.

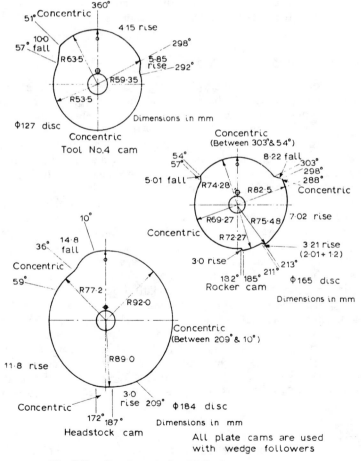

Fig 4.19 Cam Layout for Sliding Head Automatic.

as for turret auto's and in this case is equal to $\frac{374}{120}$ = 3·1 seconds.

10) *Complete Angular Displacement Column*
List number of degrees for each operation in cumulative fashion from 0 to 360°.

11) *Draw the Cams*
High point of the lobes of the cams coincides with the blank circumference, as in drawing turret auto slide cams.

The final tool layout is shown at Fig 4.18 and the three cams are drawn at Fig 4.19.

The same economic considerations apply to all types of automatics when compared to other lathes. However the sliding head automatic can be said to be of more sophisticated design than other types. It is particularly useful where accuracy, finish and concentricity are important, and if the work is also long and slender.

4.6 MULTI-SPINDLE AUTOMATICS

This type of automatic is in effect several single spindle auto's grouped together and can be used for bar work or chucking work. They may have 4, 5, 6 or 8 spindles, and the component is progressively machined to its finished shape at each spindle in turn. After each set of tools have completed machining, the clutch operated spindles automatically index. A component is completed at each indexing. The machining operations for the component must be divided approximately equally between each spindle. Therefore, the component cycle time will be equal to the longest machining operation. This is similar to the problem of *line balancing* on a conveyor fed production line where each operation carried out on a part must be of approximately the same length of time.

On some designs of multi-spindle auto's the various tool slides are operated by the most appropriate one of a series of standard cams. This, combined with the fact that tooling and camming up the auto is a long job, means that the fixed costs are relatively high. This in turn means that the break-even quantity Q_E (see Section 1.3) is high so that long production runs are required to make the machine economic. In an endeavour to lower the value of Q_E, Messrs A. C. Wickman Ltd. of Coventry have incorporated many ingenious mechanisms into their multi-spindle auto design. Quadrant and lever systems which are easily adjustable for various tool slide working strokes are incorporated instead of standard cams. Hence there are no cam costs and setting up time is reduced.

The general layout of a Wickman 5 spindle automatic is shown at Fig 4.20.

The main tool slide is a centre block which traverses upon a round slide, and obviously every tool mounted upon it must have the same feed and stroke. It is used for plain turning, drilling and reaming operations, and can carry threading attachments when required. There are two upper longitudinal slides at spindle positions 3 and 4; these slides have independent feeds and strokes and are used for the same purpose as the centre block. The four independent cross slides are used for the plunge cutting type of operation such as forming, chamfering, etc. When required, a cut off slide is provided at spindle position 5 at the front of the machine.

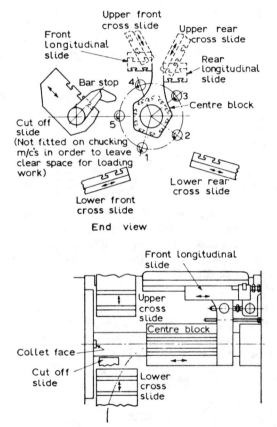

Fig 4.20   Five Spindle Automatic.

Many attachments are provided, and provision is made at spindle positions 3 and 4 for driving them. Threading is most conveniently done using collapsing taps or expanding die heads by the *differential principle* which is explained in Section 4.5. Second operation work can often be avoided by using generating attachments which can turn single flats, squares or polygons all having approximately flat faces. These attachments employ the principle in which a cutter with half the number of sides to be produced on the work is revolved in the same direction at twice the workpiece speed, the combination of these factors producing flat sides. Fig 4.21 shows an example.

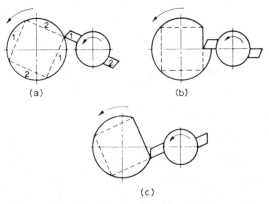

(a)   (b)

(c)

Fig 4.21   Generation of a Square.

An operator is required only to feed bar stock, castings or forgings to the machine. In the case of the chucking machine, automatic handling can be incorporated making the system fully automatic.

We will now show a simplified sequence of operations and tool position layout by courtesy of A. C. Wickman Ltd., Coventry which will illustrate the use of the tool slides.

Example 4.5

The component shown in Fig 4.22 is to be machined upon a 5 spindle automatic. Give a sequence of operations and show the relationship of the tools to each other.

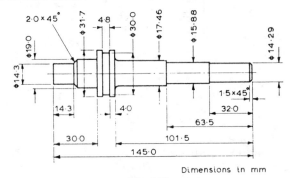

Dimensions in mm

Fig 4.22

Solution

Operation No.	Operation	Tools Used	Tool Slide Used	Spindle Position
1	Rough form 19·05 mm dia. Turn 14·29 mm dia.	Circular form tool Roller steady box	Lower front, cross slide Centre block	1
2	Finish form 19·05 mm dia. Turn 15·88 mm dia.	Circular form tool Roller steady box	Lower rear, cross slide Centre block	2
3	Form 30·96 mm dia. Turn 17·46 mm dia.	Flat form tool Roller steady box	Upper rear, cross slide Centre block	3
4	Face and chamfer end Semi-part off and break down	Roller ending box Flat tool	Centre block Upper front, cross slide	4
5	Part off Feed bar to stop	Parting off tool Cam operated Bar stop	Cut off, cross slide	5

Fig 4.23 Sequence of Operations for component being machined upon a Five Spindle Automatic.

If cylindrical grinding operations were to follow, they could be done upon a centreless grinder. Alternatively, the components could be faced and centred upon a special production machine for this purpose, followed by cylindrical grinding between centres. Fig 4.24 shows the tooling sequence.

4.7 INTERNAL BROACHING

Internal broaching is an unusual and interesting quantity production process. Metal is removed by the successive action of a number of cutting teeth incorporated in a tool called a broach. The workpiece is located and held (not usually clamped) in the machine, and the broach is either pushed or pulled through a previously rough machined hole in the workpiece. Machines may be either mechanically or hydraulically operated, and of horizontal or vertical configuration. An operator is required to thread the broach through the work and couple it to the machine ram (for pull broaching), but the process is still a fast way of producing accurate holes of any given shape. More than one broach may be required to finish an operation.

A diagram of the set-up for a pull broaching operation on a screw-operated, mechanical horizontal machine is shown at Fig 4.25.

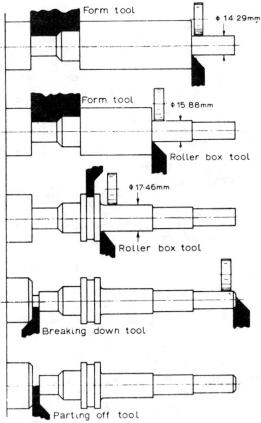

Fig 4.24 Tool Layout for Five Spindle Automatic.

Smaller broaches can be attached to the machine ram using a quick release adaptor as shown, but large broaches can be coupled using a simple cotter. The broach, which is most usually made from 18% tungsten high speed steel, may be of any profile in cross section, e.g., spline or internal gear. Accurate internal keyways in quantity are best produced by internal broaching, the set up being as shown in Fig 4.26.

The advantages of internal broaching are that very accurate holes, of any profile can be quickly produced with a high degree of surface finish. A broach can generally produce more parts between tooth regrinding than any other type of cutting tool because the cutting force per tooth is low. The broach has a long life because the first sizing tooth can be ground to the size of the last roughing tooth at each tooth regrind, hence prolonging the life of the broach considerably.

Fixed costs are comparatively high because a broach is an expensive tool to produce. Because they are invariably long and slender (push

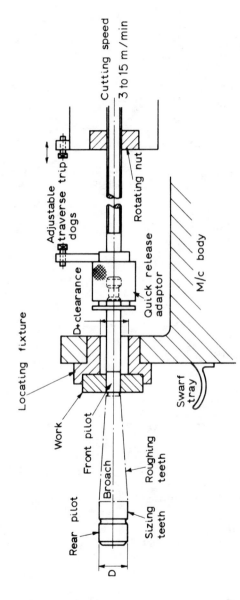

Fig 4.25 Horizontal Broaching Set-up.

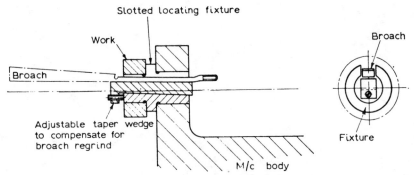

Fig 4.26   Keyway Broaching Set-up.

broaches should be as short as possible) hardening and grinding are very difficult to accomplish without distortion.

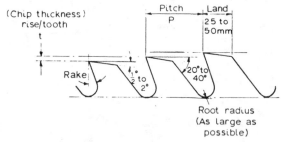

Fig 4.27(a)   Broach Roughing (Cutting) Teeth.

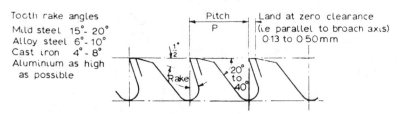

Fig 4.27(b)   Broach Finishing (Sizing) Teeth.

Broach Design

We will briefly consider the factors involved in designing an internal broach.

Tooth Shape. This is shown for roughing and finishing teeth at Fig 4.27 (a) and (b).

Burnishing teeth, which are non-cutting teeth, are occasionally included in an attempt to improve the surface finish.

Tooth Pitch (P). As a general rule the pitch should be such that at least two (preferably three) teeth are cutting at any one time. The following empirical formulae will be found satisfactory:

$$\text{Pitch } P = 1{\cdot}77 \ \sqrt{\text{length of hole}}$$

Very thin workpieces will be better clamped together and broached as a solid piece.

Rise/tooth(t). Depends upon:

a) Shape of hole.

b) Type of material being broached.

c) Force available at the machine.

d) Size of hole.

'*t*' is generally quite small, of the order of 0·025 mm to 0·160 mm.

Length of cutting portion of broach.(L).

$$L = (P \times \frac{T}{t}) + (P \times S)$$

$$= P\,(\frac{T}{t} + S) \quad \text{where } T = \text{Metal to be removed by roughing teeth.}$$

$$S = \text{Number of finishing teeth (4 to 6).}$$

Note that $\frac{T}{t}$ represents the number of roughing teeth and should be a whole number.

Load on Broach. (F). Generally $F = $ (Area of metal removed by the teeth in contact with the work) $\times$ (Force to remove 1 mm^2 of metal at a given rise/tooth).

The latter parameter is usually designated by K and is well tabulated for every different type of metal in machine tool handbooks. It does *not* equal the shear strength of the material.

For round holes $F = $ hole circumference $\times N \times t \times K$

$$= \pi dNtK \text{ where } d = \text{finished hole diameter}$$

$$N = \text{maximum number of teeth cutting at once}$$

For square holes $F = $ hole perimeter $\times N \times t \times K$

$$= 4HNtK \qquad \text{where } H = \text{finished length of one side of square}$$

In practice, after calculating F in order to find the capacity of the machine capable of pushing or pulling the broach through the hole, it would be necessary to calculate if the broach is sufficiently strong across its weakest section to withstand F in tension or compression.

Example 4.6

The bore of an alloy steel component is to be finish broached to $31{\cdot}75^{+0{\cdot}01}$ mm diameter, the bore prior to broaching being $31{\cdot}24^{+{\cdot}05}$ mm diameter. Calculate (a) pitch of teeth, (b) length of cutting portion, and (c) force to pull broach through work, if length of hole = 25 mm, t = 0·025 mm, K = 5 000N and S = 5. Sketch the broach.

Solution

a) $P = 1{\cdot}77 \sqrt{\text{hole length}} = 1{\cdot}77 \sqrt{25} = 8{\cdot}85$ mm say 9 mm
Check. This gives a minimum of two teeth cutting in the worst condition, therefore 9 mm is satisfactory.

b) $L = P(\dfrac{T}{t} + S)$ where $T = \dfrac{31{\cdot}76 - 31{\cdot}24}{2} = \dfrac{0{\cdot}52}{2}$

$$= 0{\cdot}26 \text{ mm}$$

$$\therefore L = 9\left(\frac{0{\cdot}26}{0{\cdot}025} + 5\right) = 9 \text{ (say } 11 + 5)$$

$$= 9 \times 16 \qquad = 144 \text{ mm}$$

The first cutting tooth will be 31·24 mm diameter, and the second will be 31·26 mm diameter. The rest of the cutting teeth will rise in steps of 0·05 mm on diameter, and the finishing teeth will be 31·76 mm diameter.

In practice up to 0·05 mm on diameter might be allowed for hole shrinkage after broaching.

c) $F = \pi d N t K$
                $= 3{\cdot}142 \times 31{\cdot}76 \times 3 \times 0{\cdot}025 \times 5\,000$
                $= 37{\cdot}40\, kN$
        Choose a 40 kN capacity machine.

A sketch of the broach is shown at Fig 4.28.

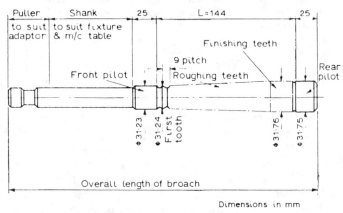

Fig 4.28   Sketch of Internal Broach.

4.8 EXTERNAL BROACHING

The broaching of external work surfaces is known as *surface broaching*, and surface broaching machines are usually vertical, and hydraulically operated. The process is similar to that of internal broaching, but the broaching force will not hold the work in position on the fixture but will push it away. Therefore, surface broaching fixtures are more elaborate requiring clamping arrangements. The process is a direct alternative to milling, and therefore the fixtures will be similar in principle to milling fixtures. (See Sections 3.6 and 3.12.)

Fig 4.29 is a diagram showing the set-up for surface broaching.

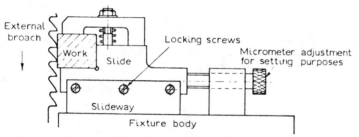

Fig 4.29 Surface Broaching.

Surface broaching is often carried out on relatively large components, and rise/tooth 't' may be as high as 0·25 mm or more, and broaches may be stellite or tungsten carbide tipped.

The following points, in addition to those already covered under internal broaching, should be considered for surface broaching:

a) Fixtures should be rigid, quick acting with fine adjustment for positioning work relative to broach cutting edges built in to fixture. Automatic operation for long runs should be considered, or a duplex (double slide) machine used with two fixtures. In the latter case one fixture will be loaded and unloaded as the other is in use. To avoid idle time the unloading and loading time should be less than the cutting time.

b) The work and fixture should have no obstructions in the plane of the surface to be broached.

c) The work and clamping must be strong enough to withstand the eccentric forces placed upon them during cutting.

d) In theory any shape of external surface can be broached, but in some cases two operations may be required in order to simplify the broach design. Such a case is shown at Fig 4.30.

It will be noticed that the broach is made up in segments, inserted and locked into position in a holder. This facilitates manufacture and re-grinding, and also means that a broken broach need not be scrapped. A progressive cutting action can be given to the broach teeth by inclining

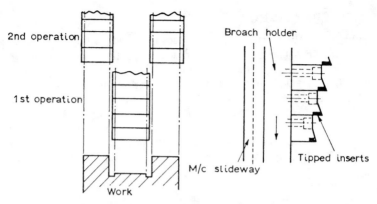

Fig 4.30 Two operation Surface Broach.

the cutting edge at an angle to the direction of travel. This is comparable to using a helical slab milling cutter as opposed to a straight fluted milling cutter.

4.9 EXTERNAL CENTRELESS GRINDING

This is the type of external cylindrical grinding process which is used for quantity production, and also for work which cannot be held between centres.

Fig 4.31 shows the principle of operation.

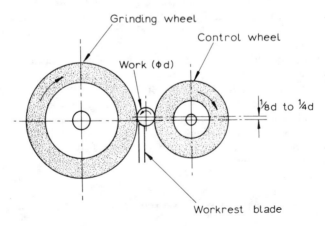

Fig 4.31 External Centreless Grinding.

The system consists of three elements:
1) *Grinding wheel*. This is the normal type of plain, straight wheel which grinds on the outside face as used for the more orthodox cylindrical

grinding of work between centres. It rotates at a peripheral speed of 25 – 30 m/s. As viewed in Fig 4.31 it rotates in a clockwise direction throwing the stream of sparks downwards.

2) *Control wheel*. It is on the same horizontal centre line as the grinding wheel. Again it is an orthodox, straight abrasive wheel but of rubber bond. It regulates the rotary action of the work, driving it as if they were two friction wheels. The peripheral speed of the work will be that of the control wheel, if there is no slip. The control wheel peripheral speed is variable between say 0·30 – 1 m/s to suit the work, and it rotates in the same direction as the grinding wheel. Hence the work rotates in an anticlockwise direction as viewed in Fig 4.31.

3) *Work rest*. This incorporates work guides for through feed grinding, and a work blade (often stellite faced to resist wear) which has a 30° angular top face to keep work against control wheel face. It is set at such a height as to keep the work axis one-eighth to one-quarter of the work diameter above centre.

There are three variations of the process:

a) *Through feed*. Used for parallel work of any length which has no surface obstructions such that it can be passed completely between the wheels set at a fixed distance apart to give the correct work diameter.

The principle of through feed grinding is shown at Fig 4.32.

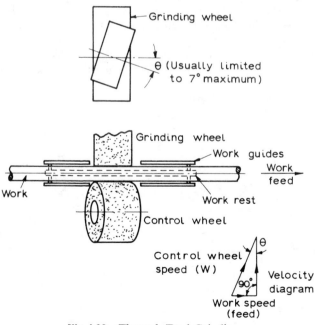

Fig 4.32 Through Feed Grinding.

As can be seen from this diagram, the control wheel is tilted through angle θ in order to impart axial feed to the work.

From the velocity diagram at Fig 4.32 the work speed $= W \sin \theta$

$\therefore$ the work feed $F = \pi dN \sin \theta$

where $d =$ control wheel diameter (mm)

$N =$ control wheel speed (rev/s)

It can also be seen from Fig 4.32 that the setting of the work guides relative to the work axis is important if the work is to be ground straight. Also, the work height above the wheels must be carefully controlled as there may be a tendency for lobing of the work, i.e., a cross section of the work being a constant diameter figure instead of a truly

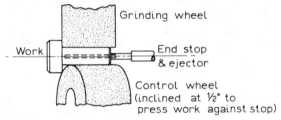

Fig 4.33   Infeed Grinding.

geometric round figure. This must always be a possibility when the work is not generated between the location of fixed, axial centres. Through feed grinding lends itself to automatic control, i.e., automatic work feed and wheel dressing.

b) *Infeed.* Corresponds to plunge cut grinding and is used for multi-diameter work, or headed work which cannot be passed completely through the wheels.

The principle of infeed grinding is shown at Fig 4.33.

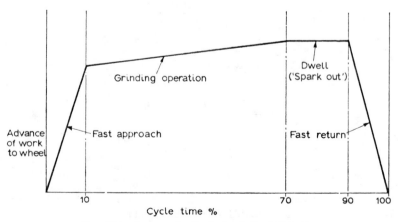

Fig 4.34   Automatic Cycle for Infeed Grinding.

The work is placed against a pre-set end stop and the control wheel slide advances the rotating work up to the grinding wheel. When the slide meets a stop the grinding wheel 'sparks out' to leave the job at the correct diameter. The slide (and hence work) retracts and the work is then automatically ejected by the end stop.

This cycle may be operator controlled but again lends itself to automatic control. This is usually done by hydraulic means. Fig 4.34 shows the timing cycle.

c) *Endfeed*. This is sometimes called the *plunge and run* technique and is used for headed work having a shank length which is too long for the wheel face, hence preventing the use of infeed grinding.

The principle of endfeed grinding is shown at Fig 4.35.

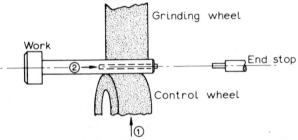

Fig 4.35 End Feed Grinding.

This is in effect a process which is a mixture of *through feed* and *infeed*. The work is plunge ground to size firstly, followed secondly by a through feed to an end stop which grinds the whole shank diameter to size. Alternatively simple endfeed grinding can be carried out for taper work without any plunge grinding being done.

Up to 150 mm diameter work can be centreless ground to possible accuracies of $\pm 0{\cdot}002\,5$ mm, from washer thickness up to 5 m long bars. Plain, headed, multi-diameter, tapered and formed work can be ground to these fine limits at very fast rates of production.

4.10 INTERNAL CENTRELESS GRINDING

This process which is not so common as external centreless grinding requires the outside diameter first to be ground, because the *internal surface* is generated from the external surface.

A diagram of the process is shown at Fig 4.36.

Again the peripheral speeds of the control wheel and work are the same, and they rotate together as a pair of friction wheels. The control wheel also supports the work, in addition to the support roll and pressure roll. After the bore has been ground, the grinding wheel retracts, followed by retraction of the pressure roll. The arm under the work ejects it. (It also lowers the next component into position if automatic loading from a hopper is used.) The pressure roll then closes up to the

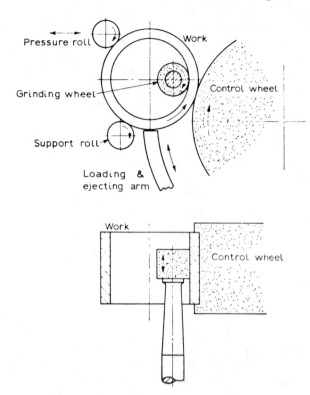

Fig 4.36 Internal Centreless Grinding.

work again. On some machines the grinding wheel spindle reciprocates
as the wheel head is mounted upon slideways; on other versions the
work reciprocates past a fixed position wheel spindle, the whole of the
work control roll unit being mounted upon slideways.

The process lends itself very neatly to automatic control, as the work
can be loaded automatically, and if required the bore size can be gauged
and controlled automatically.

The advantages are:

a) High degree of concentricity between bore and O/D.

b) High production rate, particularly where automatic loading is used.
Although fixed costs of such an installation are high, variable costs are
low as one operator can often 'mind' several machines.

c) No work holding devices required. Hence little deflection, and high
degree of accuracy is possible even with thin walled components which
otherwise could be troublesome.

d) Setting comparatively simple.

4.11 MECHANICAL COPYING SYSTEMS

Many copying systems are available for all types of machine tools, some built as an integral part of a machine, others being fitted as an attachment. There is such a variety of types that the subject warrants a text book alone, and they have been developed extensively in recent years because of their great potential as cost savers on complex formed work required in large quantities. The main requirement of any machine tool copying system is that a form on a workpiece will faithfully and consistently be reproduced from either a larger or smaller template, within the limits of accuracy required.

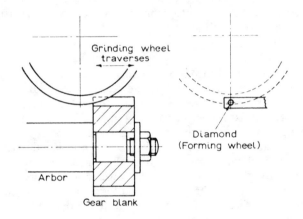

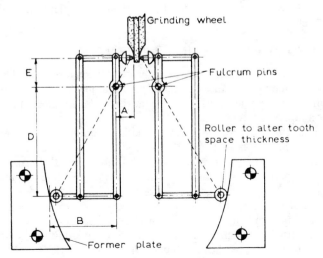

Fig 4.37 Pantograph Copying System for Gear Form Grinding Wheel.

The many systems developed make use of mechanical, hydraulic, pneumatic or electric power, and also combinations of these such as electrohydraulic. We will consider the principle of operation of three types in the last sections of this chapter, the example in this section being a mechanical type. The example selected is based upon the principle of the pantograph and is used to copy the form of an involute gear tooth space on to the grinding wheel of a gear tooth grinding machine. It is shown in Fig 4.37.

In this example, the form of the template is not of course copied directly onto the work, but is copied on the work via the grinding wheel, the accuracy being quite adequate. It will be noticed from Fig 4.37 that the follower, fulcrum and tracer must all lie on a straight line (shown dotted). Also the type of pantograph shown produces a reverse image of the template form on the wheel.

$$\text{The pantograph ratio} = \frac{D}{E} = \frac{B}{A}$$

This ratio varies usually from 5:1 to 10:1.

The correct tooth width on the ground gear is obtained using one of a series of different sized rollers with which to contact the master form plate. A dividing fixture with accurately divided plates is used for the grinding operation and is suitable for gears having up to 120 teeth. Each tooth space is ground separately, and the gear is indexed through one circular pitch after each grinding. The great advantage of the system shown is that the wheel can be lowered and re-dressed at any time without disturbing the gear and pantograph set-up. Hence wheel wear does not affect the accuracy of the finished gear. Commercial pantograph wheel forming devices are available which will accurately reproduce practically any required form from a template on to a grinding wheel.

4.12 HYDRAULIC COPYING SYSTEMS

Copying systems are used extensively on centre lathes, and a taper turning attachment is an example of a simple mechanical straight line copying device which is available for most standard lathes. Special production copying lathes are now commonplace, being used to produce the most complex forms including threads on quantity produced components. The use of expensive form tools, with a limitation on the length of their cutting edge, is eliminated. It is claimed that copy turning lathes are preferable to multi-tool lathes (described in Section 4.2) on the grounds of greater accuracy, less work deflection and less power required at the spindle.

Many ingenious hydraulic systems have been designed for lathe tracer controlled copying units, all having the advantage of having very little contact pressure between the stylus and template. (Note that it is usually possible to use a turned component as a template or copy if

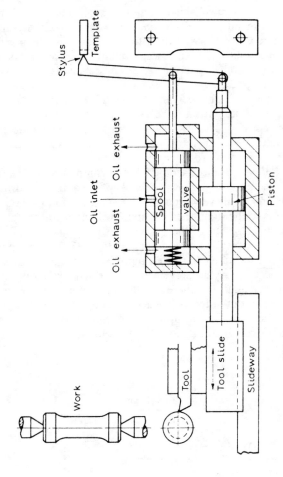

Fig 4.38 Hydraulic Servo Copying System for a Lathe.

required.) These hydraulic units are basically *servomechanisms,* i.e., a control system which magnifies a relatively small input force or signal in order to provide a larger output force or signal for operating the mechanism. This output signal from the servo mechanism must be continually and automatically modified to suit variations in the input signal, thus we then have an automatic control system. It might be thought that a simple pilot valve controlling the stroke of a hydraulic cylinder is a servomechanism, but it is not. The hand (small input force) opens the pilot valve allowing oil to one end of the cylinder giving movement of a ram (large output force). However, the position of the ram cannot be controlled and monitored precisely by means of the pilot valve once the ram is moving. A servomechanism can however give this precise degree of control.

Fig 4.38 shows in simple form the principle of operation of a hydraulic servo copying system for a lathe.

Modern copying systems designed on this principle do not have a linkage system between the piston and spool as shown, and the valve and hydraulic system are more complex.

In Fig 4.38, as the lathe saddle traverses along the bed, the stylus will follow the template edge always being kept in contact by spring pressure on the LH end of the valve spool. If the stylus, in following the template form, moves to the right, then the spool will move to the right. This will allow the oil to the left of the piston to exhaust, and will allow oil in to the right of the piston hence moving it to the left. Therefore, the tool slide and tool will move to the left reproducing the template shape upon the workpiece. A stylus movement to the left produces the opposite effect. It is essential that there is immediate tool response in answer to stylus movement in order to give accurate copying of the template. Also the hydraulic force acting upon the piston must be great enough to overcome the radial cutting force on the tool.

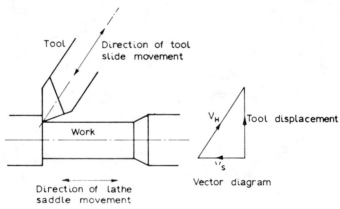

Fig 4.39

The tool slide may be set at an angle to the work axis in order to allow square shoulders to be reproduced on the work. This is shown at Fig 4.39.

To produce the square shoulder shown, the saddle traverses from right to left, and the tool slide traverses back away from the work.

In the vector diagram V_s = Saddle velocity (feed)

$\qquad\qquad\qquad V_H$ = Tool slide velocity (feed)

$\qquad\qquad\qquad \theta$ = Tool slide angle.

If the ratio $\dfrac{V_s}{V_H} = \frac{1}{2}$ and $\theta = 30°$, the tool point will be displaced along a straight line perpendicular to the work axis.

The responsiveness and accuracy of the copying system depend upon the accuracy with which the spool valve is manufactured. The spool shoulder lengths and port openings must be of a precise length, and the resistance to oil flow must be identical at any spool position.

4.13 ELECTRIC COPYING SYSTEMS

These are used extensively but require a high degree of skilled maintenance for the circuits. Mostly they use magnetic clutches with a 'make' and 'break' action to traverse the machine slides in the direction required. Electric systems are used for copying on lathes, but we will consider here their use on profile or die sinking milling machines. These are used in toolrooms for the production of metal dies and moulds required for thermo plastic moulding, pressure die casting, sheet metal drawing, drop forging, etc. An electronically coupled tracer is traversed in lines along the template (or three dimensional model) causing the end milling cutter of appropriate shape to mill the profile out of a metal die block. Increasing increments of feed are applied by the operator until the profile is completed to the correct depth.

The well known Keller system uses this principle of die sinking although the manner in which it is applied to their various types of machines is different. Fig 4.40 illustrates in simplified form the principle of operation of an electric copying system for a die sinking milling machine.

The tracing stylus is held centrally in a universal ball joint 'A' and is free to move in any direction. The back end of the stylus is centred in a cup which is an integral part of the switch lever 'B'. When the stylus is fully extended the contacts at 'C' will be closed thus completing the 'IN' circuit and operating the IN relay. As soon as the stylus moves inwards a very small amount, the contacts at 'D' will close thus completing the 'OUT' circuit and operating the OUT relay. Hence magnetic clutch 'F' will be operated which moves the machine slide and cutter head along its slideway in a direction to correspond with the stylus movement.

As the stylus is traversed, say in a vertical plane, up and down the template, it will move in and out to correspond with the profile. Either clutch 'E' or clutch 'F' will be energised causing the cutter to move in the appropriate direction. The 14 volt circuit has an earthed negative and since switch lever 'B' is also earthed, closing of either contacts 'C' or 'D' will cause a relay to be energised hence operating a magnetic clutch. Three dimensional work will obviously require a more complex system than two dimensional work.

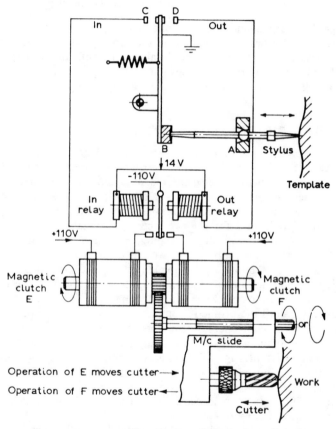

Fig 4.40 Electrical Copying System.

In the Keller system the action of the stylus is so light and sensitive that a force of approximately 1N only is required to operate the controls. A movement of less than 0·025 mm in the stylus movement is sufficient to close contacts 'C' or 'D' and cause a change in the direction of the cutter travel. The model or template can be made from any suitable material, and such is the sensitivity of the system that the grain pattern on a wooden model will be reproduced upon the metal copy.

Exercises 4

1. A 30 mm diameter mild steel bar is being turned at a spindle speed of 8 rev/s, on a multi-tool centre lathe having eight tools in operation. The feed is 0·3 mm/rev, the specific cutting pressure/tool is 2 000N/mm² and the total power consumed for the operation is 14 kW.

If the depth of cut is to be kept the same for each tool, calculate the depth of cut/tool in mm.

(Ans. 3·8 mm)

2. The component shown in Fig 4.41 is required to be produced in large quantities upon a turret type, single spindle automatic. Using a spindle speed of 30 rev/s, and feeds of 0·15 mm/rev for plain turning and drilling, 0·05 mm/rev for forming and parting off, estimate the cycle time required per component. Sketch the tool layout, produce a cam design sheet and sketch the cams.

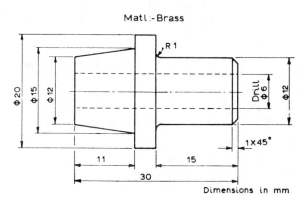

Matl:-Brass

Dimensions in mm

Fig 4.41

Assume that any required set of cycle time change gears are available so that no correcting of revolutions is necessary. Overlap operations where possible. The backshaft rotates at a constant speed of 2 rev/s 20 mm diameter bar is available.

(Ans. A cycle time of between 15 to 20 seconds should be possible depending upon sequence of operations chosen)

3. The component shown in Fig 4.42 is required to be produced in large quantities upon a single spindle sliding head automatic having four tool positions. Choosing any suitable spindle speed and feeds, and

Matl:-Free cutting steel

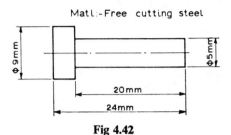

Fig 4.42

assuming that any required cycle time change gears are available draw up a cam design sheet, sketch the cams and also the tool layout 9 mm dia. bar is available.

4. A component can be produced in sufficient quantities either upon five single spindle auto's or upon one five spindle auto. Using your own figures and example of a component, compare the economics of the two processes and state which you think would be the best proposition.

5. A component having a bore of 50 mm diameter × 75 mm long is to have a keyway broached in it 12·5 mm wide × 6·4 mm deep on the bore centre line. Calculate (a) broach tooth pitch, (b) length of cutting portion of broach and (c) force required to pull the broach through the work, if rise/tooth = 0·1 mm, number of sizing teeth = 6, and $K = 3\,000$N. Sketch the broach which is to be used upon a horizontal machine, and also sketch a quick release adaptor suitable for coupling the broach to the machine ram.

(Ans. (a) 16 mm (b) 1·12 m (c) 18·75 kN)

6. Compare the process of surface broaching with those of milling and shaping, stating the advantages and disadvantages of each as a method of metal removal.

7. Plain steel spindles of 25 mm dia. × 125 mm long are to be centreless ground using the through feed method. The grinding wheel is 150 mm wide, and the control wheel which is set at 5° to the work axis is 250 mm diameter. The peripheral work speed is 0·25 m/s. Assuming no slip between the work and control wheel, calculate (a) the control wheel speed in rev/s and (b) the cycle time/piece in seconds.

(Ans. (a) 0·31 rev/s (b) 12·7)

8. With the aid of sketches compare and contrast the large scale production of multi-diameter workpieces by means of (i) a multi-tool centre lathe and (ii) a production copying lathe.

9. With the aid of sketches compare an automatic copying device operated (a) by hydraulic means and (b) by electrical means. State the advantages and disadvantages of each.

10. A 40 mm long steel bush having a bore of 35 mm and an external diameter of 55 mm can have the bore finished by internal centreless grinding, orthodox internal grinding or by internal broaching. For each process state the important factors which would govern your choice.

Further Reading

1) Browne J. W. *The Theory of Machine Tools*. Cassell and Co. Ltd.
2) Chapman W. A. J. *Workshop Technology, Part* 3. Edward Arnold Ltd.
3) Baker H. W. *Modern Workshop Technology*. Cleaver-Hulme Press Ltd.
4) Degarmo P. *Materials and Processes in Manufacturing*. The Macmillan Co.
5) Shaw T. R. *The Mechanisms of Machine Tools*. Frowde and Hodder and Stoughton Ltd.
6) 'Laying out Tools and designing Cams.' B.S.A. Tools Ltd., Birmingham.
7) 'Cam and Tool design.' Wickman Ltd., Coventry.
8) 'Turret Lathe Work.' Alfred Herbert Ltd., Coventry.
9) 'Centreless Grinding Data.' Churchill Machine Tools Ltd., Manchester.
10) De Barr A. E. 'Automatically-controlled Machine Tools.' *Journal of the Institute of Mechanical Engineers*, Jan. 1973.
11) Perkins L. J., 'The applications of special attachments on multi-spindle automatics. *Journal of the Institute of Production Engineers*, July/Aug. 1976.

CHAPTER 5

Transfer Machines

5.1 TYPES OF PRODUCTION

Before considering the principle of operation of transfer machines it would be helpful to briefly review types of production systems and the associated types of plant layout. This is done in the first two sections of this chapter.

Production systems can be grouped under three headings depending upon the quantities of parts being manufactured within the system. The types of production are (i) *Job* (ii) *Batch* (iii) *Flow*. There are no clear lines of demarcation at the boundaries of the types which in fact overlap, i.e., very small batches of two or three parts may often be produced within a jobbing system, and likewise very large batches of many thousands of parts may in effect be produced using flow production methods.

i) *Job Production*. This is the manufacture of single (one off), or very few products to a customer's individual specification. No two jobs are usually alike and orders may not always be repeated. Shipbuilding is an example of job production. The use of jigs and fixtures is not generally justified, and usually a high degree of skill and adaptability is required from the workforce and management. Production control is not easily effected, and scheduling (timetabling) and hence completion dates are often uncertain. Many small engineering jobbing shops are in existence where the skills of the shop foreman are at a premium; his role being the key one in ensuring that work progresses smoothly.

ii) *Batch Production*. This is the system of production where parts are manufactured in fixed quantities called 'batches'. The batches may be small, medium or large in number, each batch containing a number of identical articles. The determination of an optimum batch size is important, and economic factors may have to be taken into account (see Section 1.7). Batch production is the most common manufacturing system used throughout the world by metalworking industries, and is almost always associated with plant and equipment arranged in a functional layout (see Section 5.2). It is the oldest form of workplace organisation known and yet is applied increasingly, such is its utility.

When production of a batch is complete the equipment is then available for manufacture of a batch of, say different parts. Hence, general purpose machines are usually used to provide the necessary degree of flexibility. Batches of identical parts will usually pass through a series of operations in such a way that each operation is completed on the whole batch before the next operation is started. In turn, the batches may be routed upon a long and complicated path through the factory, visiting various groups of machines, before completion. It will therefore be appreciated that production planning and control may be complex and difficult, and management problems may be of a high order in batch production. In the last few years the computer has been used extensively in order to try and solve the logistical problems of this system, without noticeable success. The recent innovation of Group Technology (GT) in which similar types of products are grouped into families (see Sections 1.7 and 5.2) is an attempt to reduce some of the complications of batch production. It does seem, however, that the technique of GT has limited application.

The great advantage of batch production is the flexibility of the system which is responsive to quick changes of plan brought about by internal or external circumstances, e.g., sudden cancellation of an order. The use of general purpose machines with skilled operators and support staff allows such flexibility. Much quick and inventive jig and tool design and manufacture results from the demands of batch production. The main disadvantages of the system are high work in progress (WIP), correspondingly long throughput times, and the difficulty of obtaining accurate cost values of batches. This last factor can lead to ineffective cost control, which is part of the wider issue of inadequate production control. Batch production is practised in most engineering workshops, press shops, foundries, forges, heat-treatment shops, etc.

iii) *Flow Production.* This is the type of production system in which large quantities of identical parts are continuously manufactured and/or assembled. The fixed costs incurred in setting up such a high rate production system are large, and can only be justified by a correspondingly high rate of demand. Flow production is associated with special purpose machines, tools, fixtures, inspection devices (see Chapter 9), and handling and positioning devices such as robots. Such an automatic system lends itself to control through the medium of a computer, the term *automation* having been coined to describe it. In a flow production system the plant and equipment is arranged in a flow-line layout (see Section 5.2).

Flow production is sometimes referred to as *continuous production*, or *mass production*, both terms being descriptive of the high production rates involved. These high rates are possible because relatively complex tasks are broken down into short, simple elements. Where manual, or semiautomatic operations are carried out unskilled labour can be used. In recent years this type of production has received much adverse criticism, and is blamed for much industrial unrest due to poor job satisfaction (for the shop floor workers) and other associated difficulties. However it must be said that flow or mass production has brought low-priced standardised products, such as television sets, motor cars, etc., within easy reach of the majority of the population. Flow production can be said to create order out of dis-order (which is sometimes the state of batch production shops) and this is reflected by production planning and control being relatively simple. The biggest disadvantage is inflexibility and any change to the product can result in extensive (and expensive) changes to the equipment and layout.

Flow production is based upon the principles of *specialization, standardization* and hence *interchangeability*. Briefly these are:
a) Specialization as applied to human activities on the shop floor can be defined as the *division of labour*. This means that instead of one operator completing a whole product, he completes one small operation on the product hence becoming specialist at that one activity. This principle was known and practised during the Industrial Revolution, the technique later being perfected by Henry Ford.

As applied to flow production it means machines designed for special purposes which will complete one specific operation as efficiently as possible upon one type of component. It should ensure a high degree uniformity of work turned out.
b) Standardization. The concepts of specialization, standardization, and simplification (known as the three S's) are closely inter-related and lead to interchangeability. *Simplification* is the process of reducing the variety of products manufactured (known as *variety reduction*). Standardization is the broader concept of simplification and can be defined as the establishing of desirable criteria with respect to the size, shape, quality, etc, of a product, and if practised will lead to simplification of a company's products. This in turn will make flow production installations economically viable if the demand for the standardized product is sustained. The responsibility upon the designer in such a situation does not need emphasizing.
c) Interchangeability, or interchangeable manufacture, means that any standardized parts produced can be interchanged such that any component will assemble equally well with any mating part without any fitting being necessary. The parts produced must be as near identical as possible, and this will best be achieved where human control of the machines has been eliminated, i.e., where automation has been applied.

Before a system of interchangeability can be operated the permissible variation (tolerance) of each dimension must be decided, and also the class of fit of the mating parts must be decided. This is a cost problem as well as an engineering problem, because if the tolerances and fits are designed unnecessarily too close, this will increase the costs of manufacture. A comprehensive system of limits and fits is necessary such as BS 4 500: 1969 which is based upon the ISO recommendation R286.

5.2 TYPES OF LAYOUT

There are two basic types of plant or machine layout used in manufacturing industry, these being *functional layout* and *flow-line layout* respectively. A complete factory or one section only may conform to a particular type of layout, and sometimes combinations of the two types are used. At the end of this section we will consider *cellular layout* which is a more recent variation of the basic types.

(1) Functional Layout. The principle of this type of layout is that all machines and resources of manufacture are grouped according to their similarity of functions, ie., lathes are grouped together, milling machines are grouped together, and so on. Fig 5.1 is a diagram of a functional layout.

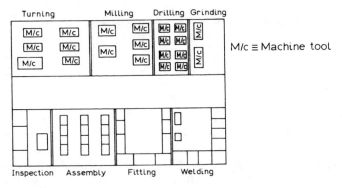

Fig 5.1 Functional Layout.

This type of layout is employed for most job and batch production. Batches of parts will travel from group to group of machine tools and equipment on a variety of routes, depending upon the product. It is common to see stocks of work in progress (WIP) being held on the shop floor in an attempt to even out the fluctuations of work flow.

The advantages of a functional layout are:
a) High degree of flexibility.
b) Uses relatively less-expensive general purpose machines.
c) Low setting-up costs.
d) Low machine tool breakdown costs.
e) Specialised supervision possible.
f) Worker job satisfaction higher.

The disadvantages are:
a) Long throughput times.
b) Planning, scheduling and control is difficult.
c) Large WIP quantities.
d) Higher degree of operative skill required.

(2) Flow-line Layout. With this type of arrangement all the machines and equipment are positioned along a flow line. The product passes from work station to work station along the flow line in a logical order according to the sequence of operations. Fig 5.2 shows the principle.

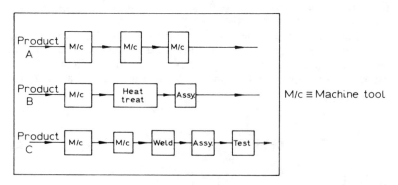

Fig 5.2   Flow-line Layout.

This type of layout is employed for flow production systems. Flow-line layouts may be simple or highly complex, and work may move in a straight line or a more complicated configuration. Again several flow lines may come together to feed a final assembly line. A car manufacturing unit probably represents the ultimate exploitation of this type of layout.

The speed at which the product is processed through the flow-line installation may be expressed as the cycle time/piece. The demand for the product will largely determine the cycle time/piece, which can be adjusted by line balancing. This is a similar problem to that discussed under multi-spindle automatics in Section 4.6. As the component travels down the line, its processing time at each station must be balanced to approximately the same time value to reduce idle time. The longest operation at any one of the stations will be the cycle time/piece. Say, for example, that the longest operation was a multi-tapping operation of 15 seconds duration, this could be reduced to $7\frac{1}{2}$ seconds by installing another multi-tapping head. The next longest operation might then be $14\frac{1}{2}$ seconds in which case the cycle time/piece has been reduced from 15 seconds to $14\frac{1}{2}$ seconds but the fixed costs have been substantially increased. The decrease in time and increase in fixed costs must be justified by the demand for the product. Normally for any flow-line layout the cycle time is chosen first. Then the layout and work stations are constructed to match the time, rather than the work being sub-divided between a pre-determined number of work stations.

The advantages of a flow-line layout are:
a) Product flows through system in logical, orderly fashion.
b) Reduced work handling leads to short cycle time/piece.
c) Less WIP.
d) Relatively simple planning and control.
e) Reduced labour skills.
f) Good space utilization.
The disadvantages are:
a) Limited flexibility.
b) Machine breakdown causes major problems.
c) High setting-up costs.
d) Uses very expensive special purpose machine.
e) Operatives work is tedious giving little job satisfaction.
Cellular layout. In the relatively modern technique of Group Technology (GT), machine tools are grouped together according to the manufacturing requirements of a family of similar components (See Section 1.7). A whole factory might be sub-divided into groups, each group performing all the operations required on a family of products. These group layouts are known as GT Cells. Usually one person is the cell leader and is responsible for its operation. Several advantages are claimed for GT systems including reduced throughput times, improved control, greater flexibility, less WIP and higher worker morale. Figure 5.3 shows a GT cell layout.

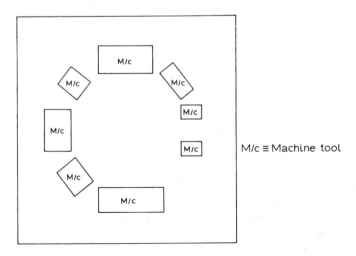

M/c ≡ Machine tool

Fig 5.3 GT Cell.

5.3 ECONOMIC JUSTIFICATION OF TRANSFER MACHINES

A transfer machine consists of several machining heads or units fastened together by conveying units, the whole constituting one, large automatic installation. Components are loaded at one end, and travel automatically past the machining heads to be successively worked on at each station, completed workpieces leaving the transfer line at the other end. Hence, the continuous production from a transfer machine may be said to have been achieved by automation, although some experts argue that true automation does not exist unless control is effected by a computer. Whilst leaving precise definition to the experts, we can agree that the installation of a transfer line can only be economically justified if the continuous production of the product is equalled by the demand for it. The fixed cost of a transfer machine may be enormous (although the variable cost will be small due to the negligible labour costs), and accurate market research will be necessary in order to ensure that future demand will not fall below a certain minimum quantity of components.

A transfer machine once installed is inflexible and can produce usually one product and one product only in large quantities. In its design, full advantage must be taken of all the aids to repetition work at high speed, such as jigs, fixtures, carbide tools, pre-set tooling, hydraulic power, special mechanical handling equipment, relays, limit switches, etc.

Transfer machines are of three basic types, which are the *in-line machine,* the *rotary indexing table machine* and the *drum machine,* these being considered separately in the next sections.

5.4 IN-LINE TRANSFER MACHINE

This consists of a straight central bed onto the sides of which are dowelled and bolted the machining heads or units at a convenient fixed pitch, usually to the order of 1 m. A diagrammatic plan view of the arrangement is shown at Fig 5.4.

The central bed could be built up to a length of several hundred metres if desired, but if floor space is not available it need not necessarily follow one straight line but could be an L, square or rectangular pattern. The component is transferred along guide rails on the central bed either with or without the use of a pallet, this depending upon the component size and shape. If pallets are used they must be returned back to the starting point, and this requires a conveyor either over, under or around the transfer machine. The work may be loaded manually or automatically onto the machine, and it is transferred by equal pitch move-

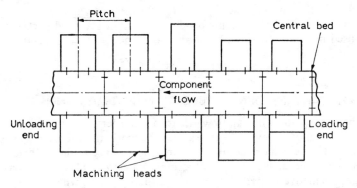

Fig 5.4 In Line Transfer Machine Arrangement.

ments from station to station where it is automatically located through dowels and clamped. Previously drilled holes will be necessary on the component, or bushed holes on the pallets. The work can be presented to the machining head in any desired position by using turntables or turn-over devices at the appropriate points on the bed.

Machining operations which can be carried out at each station include drilling, tapping, reaming, boring, counterboring, spotfacing, chamfering, countersinking, milling, grooving and sometimes even broaching. The cutting tools have a fast approach and return and may be presented to the work at any desired position or angle. Provision must be made for removing the large quantities of swarf that are produced without it being allowed to foul and clog the working parts. Coolant must be supplied and filtrated and again this is usually in large quantities. An automatic lubricating system may be required which automatically gives a shot of lubricant to the moving parts at pre-set time intervals. Automatic safety devices will be required in case of jam up of components, or a break down of some part. This means that the controls at every station must be linked so that none will operate if one is not functioning correctly.

In-line transfer machines provide the most spectacular examples of transfer machines, some being enormous in magnitude, complexity and also in terms of capital cost. The classic example is the transfer machining of cylinder blocks, and in fact other engine parts such as cylinder heads, gear box casings, and axle box casings for example are transfer machined by manufacturers of quantity produced, popular cars. The fixed cost of a transfer line can be reduced by the use of standard machining heads, standard bed parts, standard control units and switches, etc. Specialist manufacturers can provide such a range of standard units that any combination of stations required can virtually be built up without the use of special equipment. A further enormous saving is that if the machine is ever 'broken down' because of the obsolescence of the part being produced, then the standard parts can be re-used in some other combination upon a new transfer machine.

5.5 ROTARY INDEXING TABLE TRANSFER MACHINES

Where space does not allow work to be transferred in a straight line, it may be more convenient to transfer the work around a circular line. This will give a more compact arrangement hence saving floor space. This principle leads to the use of a rotary indexing table for transferring components from fixed stations of machining heads, which are spaced at equal intervals around the periphery of the table. A diagram of the arrangement is shown at Fig 5.5.

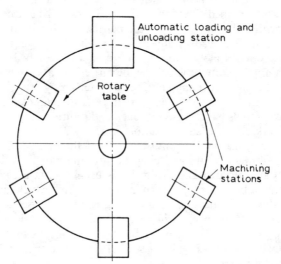

Fig 5.5   Rotary Indexing Table Transfer Machine Arrangement.

This arrangement is very like the multi-spindle automatic which was discussed in the last chapter, except that a rotary transfer machine rotates about a vertical axis, so that Fig 5.5 represents a plan view.

Much of what was said about in-line machines in the last section regarding cost, machining heads, component location, swarf and coolant control, etc., is true here, but rotary transfer machines will usually be smaller in scale. This is because there must be a limit to the size of a table which can be held and rotated in a centre base, which will be sufficiently rigid to maintain the component accuracy required. Also a further difficulty is that if a large number of component operations are required, it will clearly be impossible to accommodate them all in fixtures on a table of sensible size. This depends upon the component size, because only four stations may be possible with a large component, whereas say 16 stations may be possible with a small one.

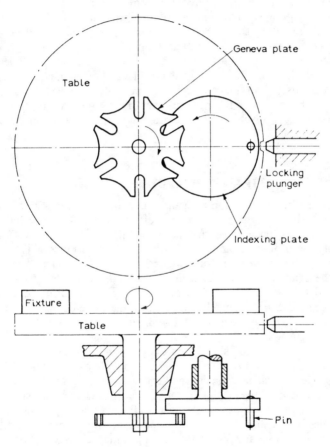

Fig 5.6 Principle of Geneva Indexing Mechanism.

An indexing rotary table should be capable of fast and precise indexing. A *geneva indexing mechanism,* similar to that used to index the circular tool turret of a turret type auto, will fulfil these conditions. The principle of such a mechanism is shown at Fig 5.6.

One revolution of the indexing plate will push the slotted geneva plate (which has six equally spaced slots) around one-sixth of a revolution by means of the pin engaging the slot. The locking plunger will then engage positively with the table to lock it in position after each indexing. The geneva plate can have any number of slots to suit the number of machining stations.

A rotary transfer machine lends itself to compactness in that the control gear can be housed in the base of the machine underneath the table. Electrically controlled hydraulic power can be used for the table operating equipment and the machining head units. A bank of limit switches controlling the various operations can be tripped in the correct sequence by pre-set cams mounted upon a shaft turning at the appropriate speed. Rotary transfer lines have been successfully installed for the complete automatic assembly of a product where no metal removal is involved. Instead of machining heads radially disposed around the table, we have presses for peening over rivets, nut running heads for assembling nuts to screws, electric brazing heads for assembly, etc.

5.6 DRUM TYPE TRANSFER MACHINES

This type of transfer machine is similar in conception to the rotary table type, but the configuration is rather different. We still have the components being transferred around a circular path to work stations radially positioned around the path at equal distances. The table, however, is replaced by a drum which is mounted upon trunnions such that the drum rotates about a horizontal axis. The work fixtures are fastened to the outside surface or periphery of the drum and where possible should be arranged to hold more than one component, as the working capacity of the machine is limited. A diagram of the layout is shown at Fig 5.7.

Reference to Fig 5.7 shows one of the limitations of the drum type machine in that the lower station(s) must be idle as it is almost always impossible to arrange a machining head to operate in the limited space under the drum. Everything in general terms which was stated for rotary transfer machines in the last section is true of drum transfer machines but the limitation of space mentioned above means that these are the least popular type of transfer machine, few having been built in this country. As with rotary machines, there is a limit to the size of drum which can conveniently be mounted and indexed upon trunnions. Therefore these machines compared to in-line transfer machines will always be small.

One last consideration which did not apply with the other two types, is that the work, at stations below the horizontal centre line, will always be hanging from the fixture. Therefore clamping must be foolproof and efficient.

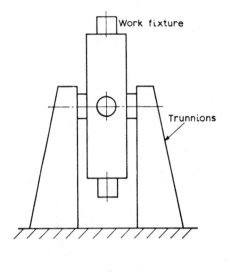

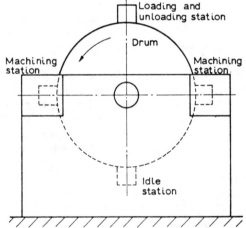

Fig 5.7 Drum Transfer Machine Arrangement.

5.7 AUTOMATIC LOADING AND TRANSFERRING METHODS

The methods which are adopted to load and transfer the work into position at each station will depend upon the type of transfer machine and the shape and size of the workpiece. In the case of large components being fed onto an in-line transfer machine, the workpieces may be placed into position at the first station by an operator using mechanical handling equipment such as electric hoists and conveyors. Alternatively

small components may be fed in large quantities into a hopper in an
indiscriminate way to be automatically loaded from the hopper into the
first station using a special hydraulic, pneumatic or electrical device.
There are many such devices operating successfully and a simple one
is shown at Fig 5.8.

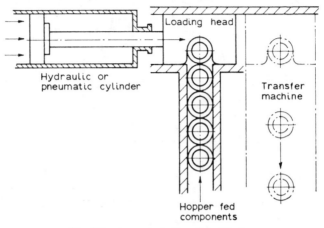

Fig 5.8   Automatic Loading Device.

The components shown are travelling down a gravity loading chute
from the hopper and are therefore queueing up ready for loading. The
component in the loading head is pushed across to drop into position
in the fixture on the transfer machine. When the loading head reverses,
the next component pushes into the head under the weight of the others
behind it. The action of the loading head cylinder is synchronised and
controlled to suit the passage of the components along (or around)
the transfer machine.

The transferring of the components accurately from station to station
can again be done in many different ways. In the case of rotary and
drum transfer machines it is a function of the indexing mechanism.
On in-line machines various mechanical devices have been used, such
as endless chain conveyors fitted with spring loaded pawls on the top.
The principle of a hydraulically operated transfer bar for an in-line
transfer machine is shown at Fig 5.9.

Reference to Fig 5.9 shows that a hydraulic cylinder is used to move
the transfer bar one work station pitch. Also hydraulic cylinders are
used to clamp the work upwards, off the guide rails, into position in the
fixture ready for machining. The transfer bar reverses, and the spring
loaded pusher dogs retract, to spring back into operating position
again at the end of the stroke. As the transfer bar moves forward again,
the dogs push the work (or work pallet) along the guide rails to the
next station. The advantage of hydraulic power is that it can be used to

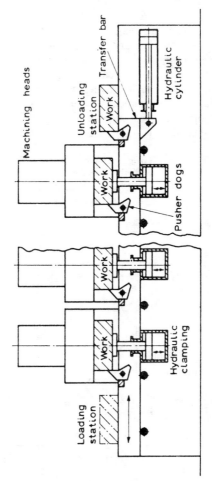

Fig 5.9 Hydraulic Transfer Mechanism.

operate not only transfer devices, but loading devices, machining heads, clamping devices, etc. Although it is expensive, it lends itself to automatic control.

5.8 MACHINING HEADS

The machining stations on transfer machines are often made up of unit heads, such as multi-spindle drilling heads for example, the spindles of which are usually driven from a self-contained electric motor through conventional reduction gearing. The heads can be mechanically or hydraulically traversed along the slideways ˙while the cutting tools work upon the component. A quick return movement is then desirable. Using mechanical means, a cam or screw can provide the medium through which the head is traversed. On modern machines the electrically controlled, hydraulically operated unit head is preferred. The principle of operation of such a head (cutting tools not shown) is outlined in Fig 5.10.

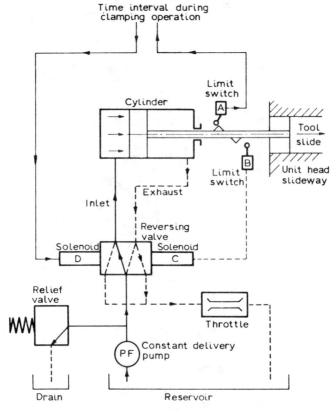

Fig 5.10   Electro-hydraulic Unit Head.

The diagram shows the tool slide stroke being actuated by a hydraulic cylinder, the direction of movement of the piston being controlled by a spool type reversing valve. This in turn is operated by solenoids C and D which push the spool across (hence reversing the direction of oil flow from the pump) when energised by spring off-set limit switches. The tool slide is moving forward, as shown in Fig 5.10, the oil from the pump impinging upon the full piston area. On the reverse stroke, the oil will impinge upon a reduced piston area (mm²) because of the piston rod. If the volume of oil (mm³/s) delivered from the pump is constant, then the speed of the reverse stroke (mm/s) must be faster than the forward stroke. On the forward stroke the speed of the tool slide movement can be controlled by passing the exhaust oil from the reversing valve through a throttle valve. This is known as *metering out,* as opposed to *metering in* where the throttle valve is placed in the inlet line. Metering out is preferred because an adjustable back pressure can be created which cushions the cutting thrust. The relief valve is provided in the inlet line to guard against over-load.

The forward movement of the tool slide is stopped when the trip dog actuates limit switch B. This in turn energises solenoid C which operates the reversing valve, hence reversing the oil flow and also the tool slide. At the end of the reverse stroke, limit switch A is actuated by a trip dog and it will stay so until the piston moves onto its forward stroke. Switch A activates, through solenoid operation, the hydraulic circuit for un-clamping and re-clamping the workpiece by means of a hydraulic cylinder. A trip dog on this cylinder will eventually operate a limit switch which energises solenoid D, hence operating the reversing valve and moving the tool slide forward on its cutting stroke. On some installations the pump (and electric motor driving it) are separate from the machining unit heads making one large driving and control unit. On others, the hydraulic system, including motor and pump, may be built into each head giving self contained separate units.

The advantage of hydraulic heads over mechanical heads is that the infinitely variable forward traverse speed, fast return speed, and the ability to run the slide against a stop for dead length work are more easily obtained from a hydraulic system than from a mechanical system.

5.9 AUTOMATIC INSPECTION

The subject of the automatic control of size is covered in more detail in Chapter 9. However, we might consider here its application to a transfer machine. Some degree of control is required over the accuracy of the work leaving the machine, and inspection stations are required at suitable intervals along the line. The frequency of inspection points will be decided by the type of operation being carried out at each machining station, and the degree of control that is required over dimensional accuracy. This, like so many production problems, is a question of cost, as each additional inspection station increases the fixed cost which must be justified by the quality of the final product. There must be an optimum point, as too little inspection will cause too many rejects to escape detection (and hence increase costs), and too much inspection will allow very few rejects to escape detection but at a much increased cost.

Manual inspection will generally be too slow to keep up with the continuous production coming off the machine, and hence relatively expensive automatic inspection equipment must be provided. Each component will be automatically gauged as it passes through the inspection station and the rejects automatically segregated, hence 100% inspection of any required feature being carried out, which is not possible by human operators. The control system at the inspection station can be based upon the 'closed loop' system, i.e., feed back is incorporated such that tool wear is automatically corrected as rejects are sorted out by the inspection unit; or can be based upon the 'open loop' system, i.e., no automatic feed back and monitoring of the tools is provided, but a warning light and/or the machine is stopped if more than an acceptable number of rejects is detected. Corrective action must then be supplied by a tool setter. The first of these two systems is much more complex as it is self regulating, and therefore far more expensive. Again it becomes a question of economics as to which is preferable, because the cost of stopping an expensive and complex transfer machine to re-set tools (using the open loop system) is very high.

One common automatic inspection device uses a probe (or probes) which is fed under hydraulic power to every work piece at the inspection station. The probe is set to record the depth, either of a drilled hole, or milled slot for example. If the probe records the correct depth, no action

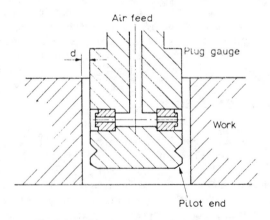

Fig 5.11   Air Gauging Hole Diameter.

takes place, but if the probe is stopped in the wrong position then the remedial action described above is activated.

Pneumatic power lends itself very neatly to automatic sizing, and air gauging has been successfully used on machines for many years. Figure 5.11 shows an air plug gauge which is being used to automatically inspect the diameter of a hole in the component.

The plug gauge shown has two jets which allow air, being fed to the gauge at constant pressure, to escape. This rate of escape of air will depend upon the clearance 'd' between the gauge and bore, and can be measured in terms of pressure. The larger the clearance the greater will be the rate of air flow from the plug gauge, and the lower will be the back pressure created in the plug at the inlet point, and vice versa. If the gauge is set to a master ring gauge of mean diameter and records a certain pressure, then variations in component hole size will cause measurable variations in this datum pressure. This again can be used to bring about remedial action. Using this type of device, means is sometimes provided for the jet of air to clean the bore prior to measurement.

Air gauging will be further considered in Chapter 9.

5.10 TOOL SERVICING

The efficiency of a transfer machine will be greatly affected by how quickly worn tools can be detected and replaced by sharp tools. If it is left to an operator to determine when a tool is worn out, and then to stop the whole machine to replace the one tool, it is going to prove a costly method of tool servicing. It is far more sensible to use a planned programme of tool replacement where batches of worn tools are replaced, after a predetermined time period, during the lunch hour or shut down period between shifts. The pre-determined time period during which the tools are used will depend upon the rate of wear of the tools, which in turn depends upon speeds used, material used, etc. An optimum cutting speed and feed is selected in practice based upon the principles outlined in Section 1.8. For full efficiency, the adoption of this system requires two essential features. Firstly, some form of counter with a dial is required which counts the number of components machined by each tool, and with the limits between which the tool should be changed marked upon the dial. Secondly, tool service stands are required near or on the machine which carry a complete set of tools pre-set in their tool holders. This latter feature reduces the cost of tool change-over to a minimum.

In practice, it may not be possible to select precise optimum speeds for minimum metal removal costs, because from the point of view of keeping machine stoppage time to a minimum, as many tools as possible should be changed at one batch. Some of course, will be on the top limit of wear, and others on the bottom limit and hence only partially worn. Somewhere between the two requirements a balance must be struck.

Broken tools are a hazard which must be considered, as considerable damage can be done at a subsequent machining station if part of a broken drill for example, is left in a hole. The hydraulic depth probes used for inspection (Section 5.9) can serve the dual purpose of inspection and detection. If the probe detects a broken drill it can give warning. An alternative way is to have each tool which might be subject to breakage surrounded by an electrical coil. If the tool breaks during cutting then the coil will not surround all the tool when it is returned to its starting point. This disturbs the characteristic of the circuit into which the coil is wired and will automatically lead to the machine being stopped.

5.11 TRANSFER PRESS

Like a transfer machine, a transfer press carries several stations and one stroke or cycle of the press ram produces one component, the component having been successively worked upon at each press station. If quantities justify the large fixed costs, a transfer press can replace several presses which might be required to produce the component, and the savings in floor space are considerable.

It is necessary to distinguish between a progressive, combined blanking and forming tool (see Section 12.6) mounted on a press, and the specially designed transfer press. With the former the metal strip is roll fed into the tool, the die consisting of one large tool with as many holes for the punches as required. It will be necessary to have an arrangement on the press tool so that the scrap strip left on top of the die is cut up and ejected off the tool. Such a progressive press tool is very difficult and expensive to design and manufacture. The transfer press, on the other hand, is fed with coil metal to the first station where the component is blanked out. This blank is automatically ejected and transferred to the next and subsequent stations until the finished component is ejected from the last station. The transfer press has provision for several separate tools to be set up between the platens, and each tool can be independently set in its own tool holder to its own shut height. Automatic grippers are attached to transfer rails running the whole press bed length. The press cycle is:

a) Ram descends and each tool works upon a component.

b) Each component is gripped ready for transfer to the next station.

c) The transfer rail moves forward one pitch (similar to that shown in Fig 5.9), and the grippers release the component into the die aperture of the next tool.

d) The transfer rails move back one pitch, and the cycle is repeated.

This cycle is carried out at the highest possible speed, giving the fastest possible production.

Figure 5.12 shows a typical component which could be produced upon a transfer press in, say nine stages.

After ejection from the final drawing tool at stage 9, the finished component will be gripped and transferred to a chute which will deposit the part into a pallet. The sequence of operations for this component illustrates the great advantage of transfer presses, viz., the component is passed from stage to stage at such great speed that the working temperature for each operation remains approximately constant. Therefore, it is not necessary to anneal the component after each drawing stage (or stages) in order to eliminate age hardening or work hardening. Intermediate annealing operations would be necessary if the component were produced on a progressive tool, or alternatively upon a series of separate presses, hence adding to the component cost.

Other variations of transfer processing are transfer welding and transfer assembly. Car bodies are spot welded together on a transfer welding line, the separate pressings being 'stitched' together by duplex spot welding machines, station by station.

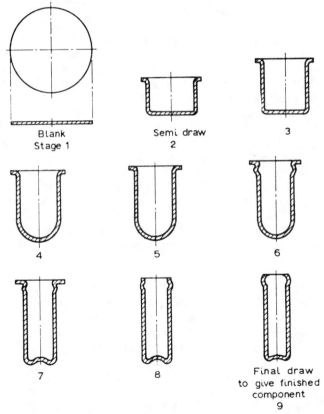

Fig 5.12 Production stages for a component being produced upon a Transfer Press.

Transfer assembly machines are used where welding may or may not be involved. The common electric light bulb, for example, is assembled together by transfer machines using many ingenious holding and transfer devices and operating at very high speeds. It is interesting to note that for years it has been feasible to build an assembly transfer machine(s) which could automatically assemble a complete car from separate component parts and sub-assemblies. This example highlights the great weakness of transfer machines, viz., the enormous capital cost of large, complex transfer machines, and the inflexibility of such a unit. The break even quantity for such an installation would be so large that the demand for the standard car produced upon it would need to be impossibly high.

5.12 LINKED LINES

A cheaper version of a transfer line can be obtained by linking together general purpose and/or special purpose machines by means of gravity roller conveyor and chutes, or powered belt conveyors. The components on a linked line have to be transferred from the conveyor and loaded into the machine, then transferred back to the conveyor to be carried to the next machine. The transferring and loading can be carried out by automatic devices, which can be replaced by an operator in case of breakdown. Such a production line is based upon flow line principles as is a transfer machine, but the linked line has the advantage of being cheaper and more flexible, but production rates will be less. The linked line can be laid out along any route, straight or curved, and many variations exist in industry today. Heavier cutting operations can be carried out on a linked line, such as surface broaching for example, which could not be conveniently fitted into a transfer machine. It will be noticed that any machine breakdown does not stop the production of the whole line, because unlike the transfer machine, each machining unit is independent. On very sophisticated layouts, it would be possible to remove any machine which has broken down and replace it with a replacement unit held by the maintenance department. This would reduce non-productive time to a minimum. There are obviously limits to how far one can carry out this policy.

Either of two principles can be adhered to when setting out and planning linked lines. Firstly, the line can be balanced as discussed in Section 5.2. Therefore, the production time at each machining station (consisting of one or more machines) is approximately the same. Secondly, the line need not be balanced but each (or some) machine works to a buffer stock of parts held between each machining station. This means that the idle time will be greatest on the fastest machine, and the output from the line is governed by the slowest machine. The volume of buffer stock must be considered very carefully paying due regard to the effect of machine breakdowns and the cost of storage. When sufficient information is known about all the factors concerned, this becomes an *operational research* problem, and a mathematical model leads to an acceptable answer. If the O.R. specialist is to construct a satisfactory model on paper, he will often have to make reasonable assumptions in order to be able to express the parameters involved mathematically. Many complex storage problems can be solved using these techniques.

Figure 5.13 shows a diagrammatic layout of a linked line.

A standard cast iron cylinder block, for example, required in very large quantities would be produced upon a transfer machine. Cylinder blocks required in smaller quantities however, could have the main machining operations carried out on a linked line, as shown in Fig 5.14, and the rest of the machining could then be done as required in batches on available machines. This gives greater flexibility.

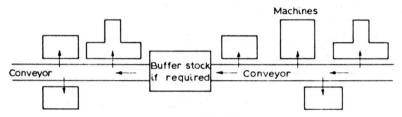

Fig 5.13   Arrangement of a Linked Line,

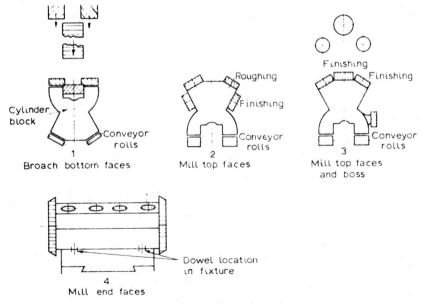

Fig 5.14   Cylinder Block Machining upon a Linked Line.

Note in Fig 5.14 that the block is transferred between machines on gravity roller conveyors, mechanical handling equipment being required for lifting. The broaching is carried out on a horizontal surface broaching machine, and the milling is carried out on fixed bed type machines. These are production type machines, which are more rigid and simple than the general purpose milling machine. Stellite tipped broaches and tungsten carbide inserted tooth face mills are used.

Exercises 5

1. (a) Distinguish between the three basic types of production, stating the advantages and limitations of each type.
   (b) Give one industrial example of the use of each type respectively.
2. (a) With the aid of sketches describe the two basic types of plant layout.
   (b) What are the major advantages and disadvantages of each type of layout?
3. Describe the technique of Group Technology (GT) and outline the type of cellular layout which might be used to implement GT.
4. (a) Sketch a simple engineering product, and set out a sequence of operations for its manufacture.
   (b) Draw (i) a functional layout, and (ii) a flow-line layout, either of which could be used for the production of the product.
5. (a) What is meant by line balancing?
   (b) A product is being manufactured upon a linked line having 12 stations in the following operational sequence:

Op. No.	Machining Station	Operation Time (Min)
1	1	3.0
2	2	1·6
3	3	4·3
4	4	2·4
5	5	3·0
6	6	2·1
7	7	1·9
8	8	1·9
9	9	1·8
10	10	4·1
11	11	0·5
12	12	2·7

The demand for the product is 1 500 pieces/week of 40 hours. Suggest line balancing which will give minimum costs/piece if the following assumptions are made:
a) Idle time on a machine costs half that of machining time.
b) Every machine installed at each station costs £1 000.
c) Storage costs/week for every piece produced over 1 500 is £1/week.
d) Machining cost/piece (not counting capital cost of machine) is 24p.

Ignore all other costs and factors in order to keep the problem relatively simple. Your answer should give the required number of machines at each station, and no book answer is given because trial and error techniques must be adopted.

6. Differentiate between (i) in-line, (ii) rotary indexing and (iii) drum type transfer machines. Give the advantages and disadvantages of each type.

7. Differentiate between transfer machines and linked lines. Using your own example and figures, give a simple cost analysis comparing the use of each system.

8. Show in outline your ideas for a preventive (planned) maintenance scheme for a large in-line transfer machine.

9. A unit machining head is required to have a (i) mechanical operated feed or (ii) a hydraulic operated feed.

With the aid of sketches show the principle of operation of each type.

10. Sketch the principle of operation of a mechanical transfer bar for an in-line machine which you think would be suitable for transferring plain blocks.

11. What is the connection between the continuous production of a transfer machine and the concepts of (a) standardization, (b) interchangeability, and (c) specialization?

12. With the aid of sketches, discuss the differences in producing a component on (a) individual presses linked by conveyors, (b) a progressive combination tool set on a single press, and (c) a transfer press.

Further Reading

1) Greene D. E. *Production Technology*. Chapman and Hall.

2) 'Payne Semblamatic Unit-Type Rotary Transfer Machines.' *Machinery*, London. No. 109, September 14th, 1966.

3) Town H. C. 'Hydraulics Applied To Unit Heads and Transfer Machines.' *Fluid Power International — No*. 31, October, 1966.

4) Schuler L. 'The Economics of Transfer Pressing.' *Journal of the Institute of Production Engineers*. October, 1959.

5) Woollard F. G. *Principles of Mass and Flow Production*. Iliffe & Sons Ltd.

6) Rathmill K., Leonard R., Davies B. J., 'Characteristics of future batch production systems.' *Journal of the Institute of Mechanical Engineers*. March 1977.

7) Rathmill K., Leonard R., Davies B. J., 'Group Technology – a restricted manufacturing philosophy.' *Journal of the Institute of Mechanical Engineers*. Dec. 1977.

8) Gallagher C. C. 'The history of batch production and functional factory layout.' *Journal of the Institute of Mechanical Engineers*. April 1980.

CHAPTER 6

Numerical Control of Machine Tools

6.1 MACHINE TOOL CONTROL

EVER SINCE machine tools have been used as a means of quantity production in the engineering industry, the trend has been to invent devices which make the machines fully automatic. This is in order to reduce the labour costs which have increasingly become the largest single cost factor in the total cost of a product. One hopes that the working labour force of this country will continue to benefit by means of increased leisure time, and where possible by employment in more interesting work. Terms like *automation* and *cybernetics* (the study of self-organising machines) have been coined as a result of the growth of control systems. The techniques have been applied to machine tools in many ways. For example, the automatic lathe discussed in Chapter 4 represents a mechanical control system, and the copying attachments also discussed in Chapter 4 represent mechanical, hydraulic or electrical control systems. The transfer machines discussed in Chapter 5 represent the pinnacle of achievement of control engineers. These complex control systems with all the paraphernalia of static switching devices used on the modern machines such as cold-cathode diodes, semi-conductors, inductive proximity switches, etc., require knowledge of hydraulic, electric and electronic techniques in order to fully understand their working details. However, we are concerned with principles, not detailed working knowledge, and it can be simply stated that any machine tool control system, no matter how small or large, is either an *open loop* or a *closed loop* system.

Open Loop Control System

If an input signal or command is given to the machine tool to carry out a certain operation, the tool may move to its ordered position and then stop. If the loop is open, then there is no *feed back* and the control system has generated no return signal which will indicate that the tool has moved to the correct position. The control loop has not been closed; there is no indication that command and result match, and consequently no means of knowing that there is an error where one exists.

Closed Loop Control System

In this case there is *feed back* built into the system. Hence, as the tool moves to its new position as the result of a command signal, its position is monitored automatically. A signal is fed back to the control unit which indicates that the tool has either moved correctly to its new position, or it has not. If not, its position is automatically corrected until it is in the right position. In general, a closed loop system will be more expensive to apply to machine tool control than an open loop system.

The turret automatic (Section 4.4) is an example of an open loop control system. Here the tools are automatically traversed to their required position (as a result of a command signal from a cam), but there is no feed back to indicate that a tool has reached its correct position.

The hydraulic copier (Section 4.12) which is a servomechanism is an example of a closed loop system. Here, the tool is re-acting continuously to input signals which are 'sensed' and fed back into the control system. Hence the copying function becomes self-regulating and does not need an operator to check or monitor the tool position and to alter it if necessary. In the case of the automatic, the work must be measured to see if the tool position is varying, and the operator will then re-set the tool position as and when required.

In this chapter we will consider the application of numerical control systems to machine tools.

6.2 THE NUMERICAL CONTROL SYSTEM

In this system, the programme from which the command signals are derived is stored in numbers, hence the term *numerical control*. This method is ideally suited for the small batch production of parts, particularly very complex parts which require great skill on the part of a machine operator. It is also suited for larger batch production (see Example 1.5), and the tapes can be stored so that the machine can be quickly set up at any time and the same programme used. Consistency and elimination of error are natural advantages which accrue.

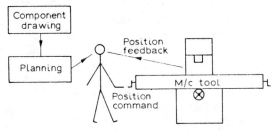

Fig 6.1 Operator Controlled Machine Tool.

In order to help understand a numerically controlled machine tool, we will look first at an operator controlled machine tool (see Fig 6.1) and then see how the human controlled operating functions are replaced by numerically controlled operating functions (see Fig 6.2).

Some initial planning takes place before the drawing and material are passed to the operator. The information is fed into the machine via hand-wheels, switches, etc. The operator interprets the component drawing in linear and angular measures which are acceptable to him and the machine. The operator monitors the cutter position during machining and makes any necessary corrections to ensure suitable output. The system is therefore closed loop with the operator carrying out data processing, position input, position feedback and compensation functions.

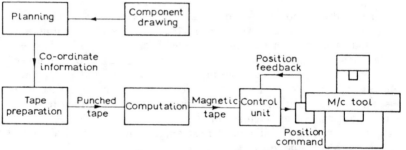

Fig 6.2   Numerically Controlled Machine Tool.

Various functions carried out by the operator (Fig 6.1) are now undertaken by other means. The machine tool is equipped with a control unit which feeds the position command information to slide-way transmission elements and compares this position command with a position feed back signal derived from automatic monitoring of the machine tool slide position. The component drawing must be translated into a form acceptable to control unit. This is the data processing part of the system where the co-ordinate information is recorded on a tape by means of a teleprinter. Again, the system is closed loop because of the feed back link.

The human operator in addition to controlling the machine has the ability to draw conclusions from the machine behaviour and act accordingly. He can compensate (through experience) for backlash in the leadscrew, slide friction, lack of stiffness, etc., and approach the final cut gradually before final commitment to it. The numerical controller on the other hand is intolerant of these characteristics and can only accept an ideal concept. The first effect of fitting numerical control therefore is to decrease accuracy of working (between 0·025 mm and 0·075 mm) compared with that which can be attained by a skilled and careful operator.

Certain aspects of machine design assume greater importance if satisfactory results are to be achieved, and great improvements have been made in overcoming friction problems with hydrostatic lubricated slideways or rollers and backlash problems with recirculating ball leadscrew and nut for example. Note that monitoring devices which feed back information only give positional information about the table (to which workpiece is fastened) relative to the slides, not about the cutting edge relative to the workpiece. The positional measuring (monitoring) devices in general usage are more than accurate enough when above limitations are considered.

Classification of NC types
There are a variety of NC systems in use, and the great majority are used on metal-cutting machine tools, particularly milling, drilling, boring and turning machines. NC grinding machines and machining centres (automatic, multi-purpose machine tools) are also used. A minority area of NC application is on flame-cutting machines, turret presses, electron-beam welding machines, etc.

Whatever the type of NC machine in use, it will have the capability of carrying out one or more of the following processes: (a) Positioning (P), (b) Line motion (L), (c) Contouring (C), in any one or more of three axes $(X, Y$ or $Z)$.

It has been found convenient to use the abbreviations shown in order to classify NC machine tools, as the following examples show: a NC vertical drilling machine with positional control to the table axes, and controlled linear motion on the vertical spindle feed would be classed 2PL; a NC vertical milling machine with contouring facilities in all three axes would be classed 3C; and a NC lathe with contouring facilities in two axes would be classed 2C.

Development of NC systems
Development of NC systems is still proceeding at a fast pace. The earlier systems (most of which are still in use) consisted of purpose-built control units permanently connected to machine tools. These 'dedicated' or 'hard-wired' systems are termed NC machines. Clearly they are relatively inflexible as they are special-purpose machine tools. Recent developments in the electronics industry particularly in the areas of miniaturisation and integration of circuits, has led to the introduction of new, small and powerful computers which can be used to control machine tools in place of the conventional dedicated controller. These newer control systems are known under the general term Computer Numerical Control (CNC). To the user the control system 'hardware' (i.e., the component parts which make up the control system) is in principle the same. The advantages of CNC are related to the control system 'software' (i.e., the information fed to, and stored in the control system) which allows a great degree of flexibility not ob-

tainable with a NC system. It is possible to extend a numerical control system such that using a single computer and data transmission lines, several machine tools can be controlled. These systems are known under the general term Direct Numerical Control (DNC). The maximum number of machine tools which can be controlled in a DNC system is dependant upon the type of machine and parts being produced. However, a minimum of 5 to 10 machine tools is considered necessary to make the system economically justifiable. In America, some DNC systems exist in which up to 256 NC machines are controlled from one computer. The most important difference (and advantage) between DNC and other NC systems is that punched tape is not used directly to control the machine tool. Instead, all information flows from a computer which interfaces with each machine control unit. Figure 6.3 shows in simplified form the differences of principle between NC, CNC and DNC systems of numerical control.

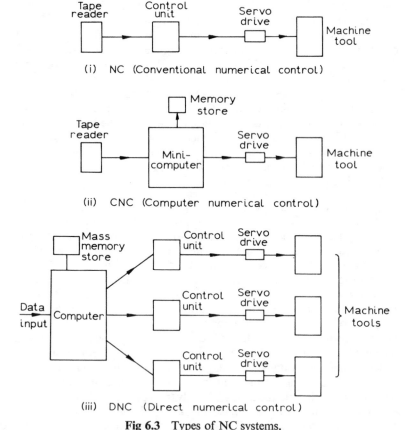

(i)   NC (Conventional numerical control)

(ii)   CNC (Computer numerical control)

(iii)   DNC (Direct numerical control)

Fig 6.3   Types of NC systems.

On all NC machines the control information required to operate the system must be translated from the part or component drawing in the form of numbers which a computer can understand. The *binary numbering system* is used for NC machines.

6.3 BINARY NUMBERS

Let us first look at the familiar *denary numerical system* used for our everyday arithmetic. This contains 10 digits which are 0, 1, 2, 3, 4, 5, 6, 7, 8, and 9 respectively, and any number used in the denary system is based upon powers of 10.

e.g., 3 406 is made up of:
$$(3 \times 10^3) + (4 \times 10^2) + (0 \times 10^1) + (6 \times 10^0)$$
$$= 3\,000 + 400 + 0 + 6$$
$$= 3\,406$$

The *binary numerical system,* on the other hand, contains two digits which are 0 and 1 respectively, and any number used in the binary system is based upon powers of 2.

e.g., 1011 is made up of:
$$(1 \times 2^3) + (0 \times 2^2) + (1 \times 2^1) + (1 \times 2^0)$$
$$= 8 + 0 + 2 + 1$$
$$= 11$$

Hence, the number 11 in the denary system can be represented by 1011 (one-nought-one-one, not one thousand and eleven) in the binary system.

Some binary numbers and their denary equivalents are given in the table shown below.

Binary Number	Denary Equivalent	Derivation
0	0	(0×2^0)
1	1	(1×2^0)
10	2	$(1 \times 2^1) + (0 \times 2^0)$
11	3	$(1 \times 2^1) + (1 \times 2^0)$
100	4	$(1 \times 2^2) + (0 \times 2^1) + (0 \times 2^0)$
111	7	$(1 \times 2^2) + (1 \times 2^1) + (1 \times 2^0)$
10000	16	
10101	21	

The last two derivations are left as an exercise.

When comparing the two systems, the binary numbers appear clumsy. For example the denary number 128 (three digits) is represented by 10000000 (eight digits) in binary form. The point is of course, that any binary number can never have more than two types of digits, i.e., 0 and 1. To the electronics engineer this represents the simplest way of representing a number by means of an electrical impulse. Either a hole in a tape allows electrical contact to be made (ON), or the absence of a hole in the tape does not allow electrical contact to be made (OFF). Therefore, a hole can represent 1, and no hole can represent 0, hence a tape can be used to transmit any required set of numbers in binary form, and relatively simple circuitry (highly complicated to most of us) can be used in the control system.

6.4 TAPES

In most numerical control machining systems, some form of tape is used for transmitting information given by the drawing to the computer, and used again to transmit data given by the computer to the machine. Two types of tape are in common use: (a) *Punched paper tape,* and (b) *Magnetic tape.* Let us consider each.

a) *Punched paper tape.* This gives a cheap and convenient means of carrying out the first part of the exercise just described, i.e., transmitting drawing information to the computer. A punched hole can represent 1 and the absence of a hole can represent 0. In use the tape is fed over brush contacts which 'sense' the hole pattern and can detect the presence of a hole or not. The brushes close the circuit through a punched hole and trigger off signal currents which cause the appropriate parts of the computer circuit to be actuated. Figure 6.4 illustrates the principle.

In addition to electro-mechanical tape readers working on the principle described, photo-electrical tape readers are used in which the tape is scanned by concentrated light rays falling upon it which detect the presence of a hole or otherwise. This type of tape reader operates at considerably faster speeds than others, – up to 1 000 characters.

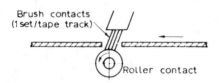

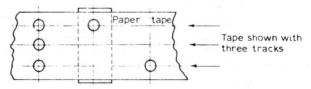

Fig 6.4   Sensing a Punched Tape.

b) *Magnetic tape.* This is usually 6 or 25 mm wide and is exactly like that used in home tape recorders. The principle is similar to that used upon a punched tape, but instead of a punched hole (1) and no punched hole (0) magnetised spots on the tape surface are used. The polarisation of the magnetic spot (either + or —) indicates the binary digit (either 1 or 0).

The feeler sensing pins of the punched tape are replaced by an electric reading head which contains an electro magnet. See Fig 6.5.

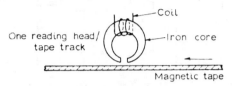

Fig 6.5 Sensing a Magnetic Tape.

Each magnetised spot on the tape which passes under the reading head will generate an electric impulse in the coil, the polarisation of the spot enabling the system to recognise 1 or 0. The big advantage of magnetic tapes is that far more information can be recorded upon them, and they can be 'read' at a speed faster than is possible with punched tape (or cards), if so desired. It is also more durable but more expensive than paper tape.

It was stated earlier that binary numbers are clumsy compared with denary numbers because of the greater number of digits required to represent a simple number. This is so, and if a true binary numbering system was used to represent one number across a row, then tapes would have to have an unwieldy number of tracks of holes (or spots) making the tape very wide.

In practice however, a modified system of binary numbers is employed, and several rows are used to represent a number. This is shown in Fig. 6.6 applied to an 8-track tape.

The first four tracks only (three to the right of the feed holes, and one to the left) are used to represent a binary number. Track 1 represents 2^0, track 2 represents 2^1, track 3 represents 2^2 and track 4 represents 2^3. Hence, across a row of tape, punched holes (or spots) placed in the appropriate position can represent any denary digit up to 9 in value.

The pattern of holes shown in Fig 6.6 represents the value 12·793 6. Any required value can be represented in the system shown by a block of six rows incorporating four tracks. With the direction of tape feed shown the hole pattern should be read from bottom to top. Therefore, numbers can be represented on the tape from 00·000 0 to 99·999 9. The decimal point is not shown on the tape because it always occurs after two digits, and its position is sensed by the computers built-in logic.

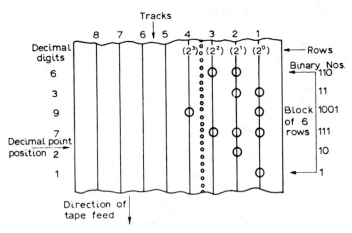

Fig 6.6 Eight Track Punched Tape.

The fifth track (in some systems) is used for *odd parity*, i.e., a hole is punched or omitted from this track such that every row in the complete tape adds up to an odd number of holes. This is a simple checking device to ensure that the tape punching and tape reading devices are functioning correctly. Should a row appear on the tape with an even number of holes, the machine using the tape will stop. The equipment must then be checked for malfunctioning.

The sixth and seventh tracks are used for what are known as *alpha characters,* i.e., alphabetical characters. These are required for coded instructions which will be explained later in the next section.

The eighth track is sometimes used to indicate the end of a block of information, i.e., one hole punched in the appropriate place in the eighth track would indicate the end of a particular sequence of information.

It should be noted that the example of a tape system cited is for an 8-track tape, but in some cases a 5-track tape, for example, is used.

There are a great variety of numerical control systems using the medium of tape control of machine tools, and this variety can be puzzling. In general, the tape control of machine tools is applied to continuous path machining, or point to point positioning. We will consider these applications separately in the next two sections.

6.5 CONTINUOUS PATH MACHINING

This covers the machining operations such as vertical profile milling, die sinking for example, and has also been successfully applied to oxy-acetylene profile cutting machines.

Figure 6.7 shows the steps involved in programming and controlling the continuous path machining of a component.

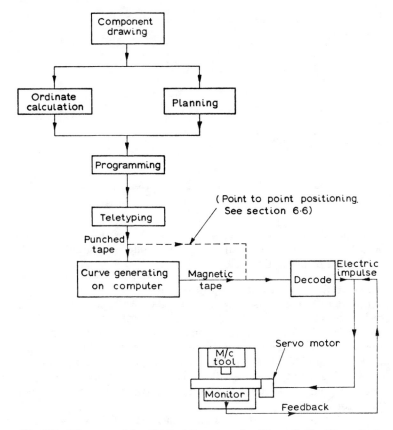

Fig 6.7 Diagram of the Control Sequence for Numerically Controlled Continuous Path Machining.

Again, there are many commercial systems available with great differences of detail, but similar in principle. We will now consider each of the steps shown in Fig 6.7.

Component Drawing

The initial component drawing produced will be of the orthodox type, suitably dimensioned. We will consider the reproduction of the component shown at Fig 6.8 which is to have the edge shown machined upon a numerical control vertical milling machine. This is machining in two planes only (i.e., X and Y planes) although in practice the tool movement can be controlled in three planes if desired, for work such as die sinking. In such a case the exercise becomes more difficult and involved.

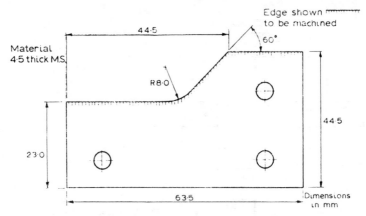

Fig 6.8 Component Drawing.

The three previously drilled holes could be used for location. In general, fixtures for numerical control machining are simpler than for other forms of machining.

Ordinate Calculating and Planning

An ordinate drawing must now be prepared on which all the dimensions are shown as rectangular co-ordinates from a datum point to which the cutter is set before commencing machining. In practice, it is only necessary to indicate the points where the profile changes (change points) with an identifying number. The change points are then written on to a table with the calculated ordinate dimensions shown alongside. Hence, the only dimensions given on the drawing are the datum point dimensions.

The ordinate drawing for the component (Fig 6.8) is shown at Fig 6.9, with the accompanying table of ordinates shown underneath the drawing.

There are one or two points worth noting from Fig 6.9. First, the cutter diameter does not have to match the component profile, as is often the case in operator controlled milling. This is because the cutter is going to track around the profile, and as long as the cutter does not 'foul' (cutter interference) the work, any sensible diameter will suffice. After the cutter has over-run the profile it is returned to the datum point by the shortest route.

Next, the direction of cutter movement is important, and the accepted convention shown at Fig 6.9 is adhered to, i.e., leftwards along the X axis is negative, and downwards along the Y axis is negative, and vice versa. The direction of movement for the example given is all positive, but it must be shown.

Finally, the co-ordinates shown are calculated from the datum to the work profile. This method is used in the Ferranti system. In the

E.M.I. system for example, the co-ordinates are calculated from the datum to the cutter centre. In either case in practice, a further ordinate would have to be given at point 0 which is the centre of the 8 mm radius, in order that the computer has sufficient information to calculate the intermediate ordinates between points three and four.

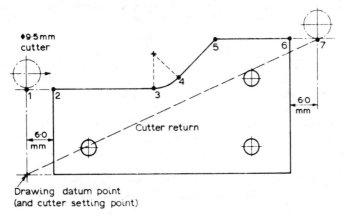

Fig 6.9   Ordinate Drawing.

Point	Co-ordinates from datum point	
	($-$) Xmm ($+$)	($-$) Ymm ($+$)
1	0	+23·0
2	+6·0	23·0
3	+34·3	23·0
4	+42·0	+27·0
5	+50·5	+44·5
6	+69·5	44·5
7	+75·5	44·5

Also at this stage, the initial planning must be carried out to do with the engineering functions of the job. Cutter diameter, cutter spindle speed, cutter feed, etc., must be decided upon.

Programming (see Section 6.8)

All the above ordinate information and planning information is written on to a planning sheet or programme, in a manner to suit the system being used. There are many different systems of programming or 'languages' in use. There are two things to consider here, viz., recording the ordinate information on the paper tape, and recording the planning information on the same tape.

The first we have already considered, and the ordinates from the programme can be shown on the tape by using the first four tape tracks in blocks, as described in Section 6.4.

The second is done by means of a code which was referred to in Section 6.4 where tape tracks 6 and 7 are used (sometimes in conjunction with tracks 1 to 4) to give a unique pattern of holes across the row to represent a function in the machining cycle. Alpha characters are used for the code; for example, say 'A' is represented by a row of holes in tracks 6 and 7, or say 'N' is represented by a row of holes in tracks 1, 2, 3, 4, 6 and 7. 'A' in the code could represent 'start cutter spindle', and 'N' in code could represent 'unclamp workpiece'. Hence, these alphabetical symbols could be written in the appropriate place in the programme to suit the machining cycle. These unique rows of punched holes (which will eventually lead to the activation of the required function on the machine) which represent coded instructions will not be used for any other purpose in the programme.

Obviously, any arbitrary grouping of symbols could be used as a code. It is economically unsound to have a variety of systems, therefore efforts are being made to standardize codes. Two codes in common use are:

(i) The BS 3635: Part 1: 1972 (Control input data) code,
(ii) EIA (Electronic Industries Association) Code, developed in the United States.

Part programming principles are developed further in section 6.8.

Teletyping

A teletypewriter (teleprinter), operated by a typist to the rules of the code, is used to punch out the pattern of holes on the tape from the information given on the programme. In other words, the tape when produced, will have the identical information punched on it in binary form, to that contained in the programme. A teleprinter is shown at Fig 6.9.

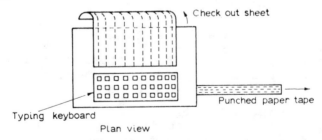

Fig 6.10 Tele-typewriter.

The keyboard has keys for the numerals and alpha characters used in the code. As the typist presses the appropriate key, the correct hole pattern will automatically be punched upon the paper tape issuing from the teleprinter. Also, a check out sheet will be fed from the front of the machine which the typist can read, and check that she is copying the programme correctly.

Curve Generating on Computer

For continuous path machining, the above programme would not contain sufficient ordinate points, because the cutter must trace out, or follow, a series of points close together, one by one around the profile. The machine in effect will be instructed to move from point to point in turn. Imagine that each point on the required profile is connected by a straight line along which the cutter axis moves between points. Then the more points which are identified by ordinates on the tape, the more accurate will be the finished profile. For example, a system required to give an accuracy of 0·01 mm would require 100 points/mm of travel. The calculations of rectangular co-ordinates for 100 points/mm would drive the average human being mad, but this is the sort of task that a computer does marvellously well at incredible speeds. If the paper tape is fed into it, the computer which understands binary language will calculate all the ordinates for the intermediate points between the change points given on the tape. In practice, the points are usually spaced 0·0025 mm apart.

The code we have already referred to in discussing the machining functions such as stop, start, clamp, change speed, etc., must be extended to cover the type of information needed by the computer during its calculations. For example, in the component shown in Fig 6.9, the computer would need to know that points 3 and 4 were on a circle of 8 mm radius. Therefore, it would require co-ordinates for the centre 0 of the radius, and also a unique pattern of holes on the tape to represent some alpha character (say 'C' for circle) which the computer was set up to recognise. Hence, it can be seen that the code will have to be fully comprehensive.

In the Ferranti system a digital computer is used which carries out the calculations described at very high speeds, then processes all the information required, including all the calculated ordinates, on to a magnetic tape of the type described under Section 6.4.

In some systems, the computer calculates all the points along the required profile as a series of straight lines 0·0025 mm apart. This is known as a *true path system,* and is really linear interpolation in the limit, and is in accordance with our ideas gained in the study of differential calculus. Another system, which is probably more logical, has the points joined by a series of parabolic curves. This is better in theory when considering the points around the curve, as a parabola can be made to approximate very closely to any curve or circular arc (or straight line). This second system is known as *parabolic interpolation.*

Decode

The magnetic tape from the computer is mounted on a play-back tape deck built into a *control unit* which is adjacent to the machine tool. This tape deck has a reading head which decodes the binary information on the tape as it feeds past it. This information is passed from the reading head to the control unit in digital form as a series of electrical pulses. The tape deck speed has a controlling influence upon the ultimate machine slide velocity along its slideway. Standard tape speeds of 381 mm/s, 190·5 mm/s and 95·25 mm/s are used.

Finally, electric impulses or signals are fed by the control unit to the servo motors which control the machine table (slide) movement and the cutter head. These command signals generated by the control unit are a function of the pulses received from the tape, but magnified in order to actuate the machine control units. All the operator has to do initially is to position the workpiece in the fixture, set the cutter to the datum point, then press the start button. Machining will then commence, and all the operations specified on the original programme will be carried out automatically until the cycle is complete.

Servo Motor

The servo motor receives and acts upon the series of electric command signals from the control unit. The motor response must be accurate to drive the work against the cutting forces.

Servo motors are usually of two types: (a) high frequency electric, or (b) hydraulic. Let us briefly examine each.

a) *High frequency electric servo motors.* These are similar in operation to the normal type of induction motors used upon machine tools at a frequency of 50 Hz. However, the servo motor is operated on a three-phase supply at 400 Hz giving full power at 167 rev/s. Consequently, the maximum power available will be limited and is usually only to the order of 0·55 kW. This is a disadvantage where reasonably heavy cutting is required.

b) *Hydraulic servo motors.* A rotary hydraulic motor is in effect the opposite of a rotary hydraulic pump. The pump rotor is driven by an electric motor, the rotation of the rotor drawing fluid into the pump and expelling it under pressure. In the case of the motor, fluid is pumped into the motor under pressure causing the rotor to turn. Hydraulic motors are used of 2·25 kW with an accuracy of response of 0·005 mm, up to 11·25 kW giving an accuracy of response of 0·025 mm.

The hydraulic fluid is metered to the motor rotor by means of an electro-hydraulic transducer which receives its command signal from the control unit. (Note. A transducer is a device used for converting a signal of one kind into a corresponding physical quantity of another kind, such as a force or displacement. A resistance strain gauge is a transducer, because a change in the length of the gauge can be related to a change in the electrical resistance of the gauge.)

If a closed loop system is required, then a table monitoring device will be necessary in order to provide feed back information relating to the table position in the form of an electrical impulse. We will deal with monitoring systems in Section 6.7.

6.6 POINT TO POINT POSITIONING

This covers the machining operations such as drilling, reaming, tapping, boring, boss face milling for example. The object of this numerical control system is to move a machine tool table (and hence the work) through rectangular co-ordinates according to a programme, in order that machining operations can be carried out at the programmed points where the table stops. The machining operation will be drilling, tapping, boring, etc. The process is similar to that involved in jig boring, and as might be expected, numerical control has now been applied to jig boring machines. Once set up the system will enable any number of components to be produced with the maximum amount of consistency of dimensional accuracy. The difference here then, compared to the system described in Section 6.5, is that no computer is required to generate the intermediate ordinates between change points on a curve.

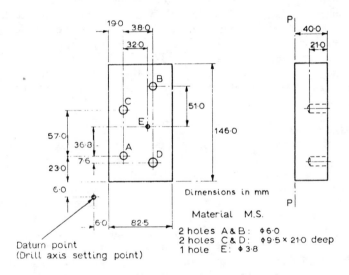

Dimensions in mm

Material M.S.

2 holes A & B: φ6·0
2 holes C & D: φ9·5 × 21·0 deep
1 hole E: φ3·8

Datum point
(Drill axis setting point)

Hole	Hole size (mm)	Co-ordinates		End of drill to face P–P	Speed (rev/s)	Feed (mm/s)
		X	Y			
A	6·0	+25·0	+29·0	Thro'	20	1·50
B	6·0	+38·0	+87·8	Thro'	20	1·50
C	9·5	−38·0	−30·8	19·0	12	2·00
D	9·5	+38·0	−64·6	19·0	12	2·00
E	3·8	−6·0	+44·4	Thro'	35	1·00

Fig 6.11 Component Drawing and Point-to-point Positioning Programme.

This is shown on the block diagram, Fig 6.7. In point to point positioning, the table will be moved any distance x along a straight line between two change points, and the ordinates can easily be programmed by a human being without the aid of a computer. Hence, the original punched tape produced by the teleprinter can be used directly to feed information to the control unit in order to send electric command signals to the table servo motors.

Consider the component shown at Fig. 6.11 which requires the holes to be drilled upon a numerically controlled vertical boring machine. The programme, ready for teleprinting upon a tape in binary numbers, is shown underneath the drawing. Had the holes been positioned by polar co-ordinates on the original design drawing, then they would have to be converted into rectangular co-ordinates as shown in Fig 6.11.

The first hole to be drilled is that nearest to the datum point. The second is of the same diameter to enable the same drill to be used. The same logical pattern of drilling is used to decide the order of drilling of any number of holes. After drilling the final hole, the machine table is returned to the starting point such that the machine spindle is positioned over the datum point. The work is then replaced by a fresh part.

The programmed information is fed to the control unit by means of the original punched tape produced by the teleprinter. In addition, *punched cards* can be used. Again, a system of *decade dials* is sometimes used for programming.

Punched Cards

These may be regarded as being identical in principle to punched paper tape. The pattern of punched holes in the card are used in exactly the same manner; as a means of recording information in binary numbers. A card could have any required number of tracks (say eight), and sufficient rows to record a block of information for the programming of one component (say 80). One hole punched in track 8 say, at the end of the block of binary information punched on the card, would signal to the control unit that the job was complete. Then the servo motors, under the action of a corresponding command signal, would return the table back to the datum position.

Punched cards are used a great deal for computer controlled systems of organization, such as production control systems, sales record systems, etc. Cards are easy to punch and store and are cheap. Today they are often made of plastic sheet which is very durable. The hole pattern can be sensed as shown at Fig 6.4 which allows electrical contacts to close in the required order. A more modern method is to use light rays and photocells. The presence of a hole in the card (or tape) is detected by a light passing through it and falling upon the appropriate photocell. The signal from the cell is amplified in order to operate a relay. (Note. A relay is a device which enables a low power current in one circuit to control a high power current in another circuit.)

Decade Dials

Instead of using tapes or cards to record the required co-ordinates, hand dials may be used which are positioned conveniently in the control unit. If settings are required in 0·01 mm increments for point to point positioning then five decade (or setting) dials could be used representing say 100 mm, 10 mm, 1 mm, 0·1 mm and 0·01 mm respectively. Each dial is calibrated to read from 0 to 9, hence any settings can be made from 000·00 mm to 999·99 mm. There are one set of dials to control the longitudinal movement of the machine table, and one set to control the crossways movement. Hence, the table can be made to move through two co-ordinate movements for one setting of the dials. They would have to be re-set, to automatically move the table to its next position. Hand setting of the type described combined with a monitoring system of high accuracy, makes this method of numerical control ideal for a jig boring machine. Continuous automatic re-setting of the decade dials can be arranged where quantity production of a component is required.

The mechanism of a decade dial setting system is complex, but in general the dials when set to the required digits control (through clutches and gearing) transmitting synchros. These in turn control the servo motor which turns the leadscrew the requisite amount. (Note. Synchro is the name given to any electromechanical device used for transmission or receiving, such as an AC position motor. This, like all rotary synchros, contains a rotor and wound stator.)

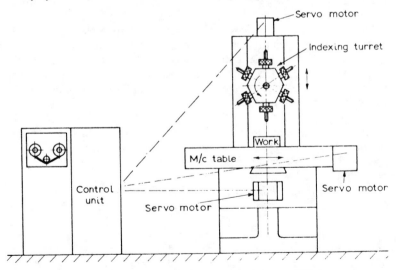

Fig 6.12 Vertical Drilling Numerical Control Unit.

Let us now return to the actual machine itself. It has been found that the greatest benefits of numerical control can be obtained by designing a complete system as a unit, rather than fitting a numerical control unit to an existing machine tool adapted with servo motors. This has meant that the newly designed machines are ergonomically and economically sound, being ideal for one specialized class of machining operations. In point to point positioning systems, the indexing turret, vertical drilling machine has evolved as the most suitable type of configuration. A diagram of this machine is shown at Fig 6.12.

Each station on the turret is set up for each operation, such as drilling, reaming, tapping, etc. Then when the tool at the station has been used the required number of times, the turret is automatically indexed to bring the next tool into position.

6.7 MONITORING SYSTEMS

The monitoring systems used for numerical control machine tools are, in effect, position transducers, which enable feed back signals relating to the table (and hence work) position to be transmitted to the control unit. All systems used are either (a) *analogue* or (b) *digital* in principle.
a) *Analogue*. This is the term used to refer to a quantity which resembles, in certain respects, another quantity. A slide rule is an analogue device, where a length on a scale is analogous (not equal to) to a number, i.e., the value 4 on the scale might be 150 mm away from the zero mark on the scale, therefore 4 can be said to be analogous to 150 mm. In analogue monitoring devices, an electrical unit such as voltage might be related to the position of work relative to the cutter axis, in mm, i.e., 10 volts, for example, could indicate 250 mm of movement of the work. All analogue feedback systems allow rapid indexing from one position to another and are ideal for point to point positioning systems.
b) *Digital*. The amount of movement of the worktable is measured in discrete (separate) quantities. The number of discrete quantities, say of 0·025 mm magnitude are counted by some device included in the monitoring system. These quantities, or digits, are counted at very high speeds. Therefore, digital feedback systems are ideal for continuous path control.

Yet again we face a complicated subject because of the great variety of position transducers in use. These are shown in the diagram at Fig 6.13.

To make matters even more complicated some systems are a mixture of analogue and digital devices, one such being indicated in the centre of the diagram. It is impossible in a book of this nature to consider all the systems, but we will consider a few which make use of quite different principles.

Firstly let us recall the three factors which are important to any measuring system, whatever the situation in which it is being used.

These are:
Accuracy which is the accordance of the measurement with the measured quantity.
Resolution which is the smallest value to which the measurement will respond.
Repeatability which is the ability of the system to repeat a given input value.

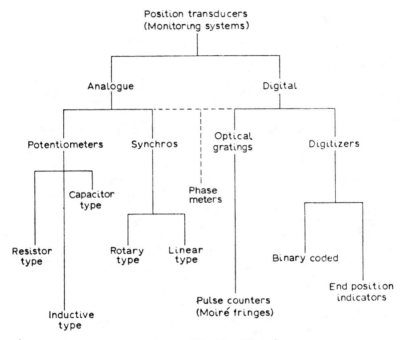

Fig 6.13   Types of Position Transducers.

Each of the systems shown in Fig 6.13 possess these three factors to a greater or lesser degree, and the choice of a monitoring system will largely depend upon which of the three factors is most important for a particular process. For example, a *resistor* (resistance potentiometer) can be used as a cheap position transducer over relatively short lengths, but its resolution is very low, i.e., it will only measure relatively coarse increments of movement which is not suitable for most machine tool applications.

Let us now look in more detail at some position transducers.

Rotary Type Synchro
This is an analogue measuring system, which is relatively simple and cheap. It measures rotation of the leadscrew, which can be related to table movement. See Fig 6.14.

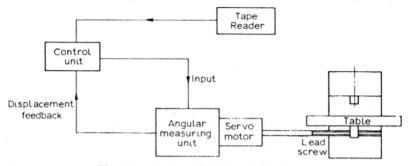

Fig 6.14   Analogue Position Transducers.

Leadscrew rotation is measured by means of scanning a potential transducer, the voltage picked off being proportional to the rotation from a machine datum. Disadvantage is that the leadscrew rotation is being measured and not the work (table) displacement, therefore variations due to mechanical linkage, and in particular backlash, are not accounted for.

This system has high accuracy if steps are taken to overcome the problems of backlash, etc., and is ideal for point to point machining.

Optical Gratings
This system may be analogue, but if combined with a pulse counting device, can be used as a digital measuring system. Although relatively expensive, it is a high class monitoring system and is worth examining in some detail. Figure 6.15 is a diagram of the main features.

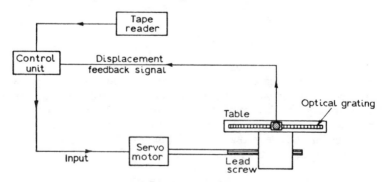

Fig 6.15   Optical Grating Monitoring System.

An optical scale or grating is fixed to the machine bed, and super-imposed above it is a short length of scale (about 19 mm diameter) fixed to the table. This small index scale is rotated at an angle to the long scale and scans the bottom scale as the table moves. These transparent gratings have surfaces which bear a large number of evenly spaced lines as grooves (say 100/mm). The grooves are impressed on to a gelatine coated glass blank, the glass acting as a backing. The grooves are too close to be numbered, but can be counted as the table moves by using two scales as described to produce moiré fringes.

Theory of Moiré Fringes

When two scales are superimposed and one is inclined to the other, illumination from the back will reveal *fringes* which run approximately at right angles to the lines of the gratings. The spacing (wavelength) of these fringes is a function of the angle θ between the lines, and the smaller the angle the larger the spacing. See Fig 6.16.

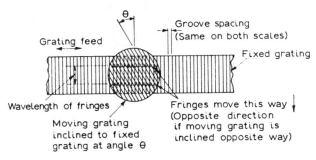

Fig 6.16   Moiré Fringes.

If the top grating is moved along as shown at right angles to the lines of the fixed grating, then a fringe will move to a position occupied by its neighbour when the top grating has traversed one groove spacing. If a limiting slot equal in width to half fringe wavelength is placed close to the top grating, as one grating is moved relative to the other, the passage of the fringes causes the illumination at the slot to fluctuate between brightness and darkness. If the eye is replaced by a light sensitive device such as a photo-cell the movement of the fringes can be electrically recorded and counted as shown in Fig 6.17. The gratings are not in contact hence no wear takes place.

Referring to Fig 6.17, the lens which is above the limiting slot collects the diffracted light (which need not be monochromatic) and focuses it at the slit. This slit is arranged to select one light of a particular wavelength from the spectra of white light, to suit the photo-cell. These types of optical gratings are sometimes called diffraction gratings because they make use of diffracted light. The latest technique is to use etched stainless steel gratings which reflect light instead of transmitting it. (Note. Monochromatic light is light of a single wavelength which requires a special light source. Diffracted light is light which is broken up into its constituent parts and is a phenomenon which can be caused by passing light through a narrow slit.)

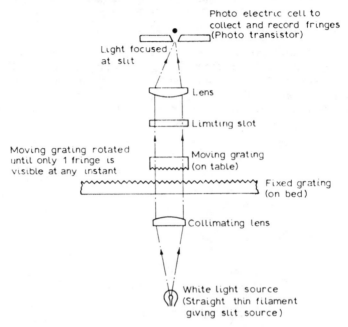

Fig 6.17   Optical System for presenting Fringes to a Photo-cell.

A pair of coarse gratings (4 lines/mm) readily demonstrate *moiré fringes,* but fine gratings used in practice require an optical system described above. The intensity of the light varies sinusoidally as one grating moves past the other, and a photo-cell placed behind the exit slit will produce an AC current, each cycle of which indicates a movement of one grating division. This current from the reading head can be dealt with in one of two ways:

a) *Digital.* The current is used to actuate an electronic pulse counter which simply registers the cycles of the current and thus counts the number of complete spaces moved by the grating. Therefore 100 lines/mm on the grating gives a resolution of 0·01 mm. The generator transmits pulses according to the number of fringes counted by the electronic counter.

b) *Analogue.* In this case the current may be fed into a phase meter and compared with a second AC current derived from another grating system. The generator includes a detector to detect variations and transmit these variations into a feed back link.

In (a) above the feedback signal specifies the table position in electrical form as a series of pulses. In (b) above the feedback signal specifies the table position as a potential difference measured in volts. This enables the command position of the table to be compared with the actual position at all times, and where error is present (i.e., feedback signal and command signal are not equal) the table position is altered to the correct position. In other words, if required, the whole system can be self correcting.

6.8 PART PROGRAMMING

The act of *programming* a *part* such that it can be machined upon an NC machine tool is too complex to describe in detail here. Hence the topic will be dealt with in general terms only. Programming must be written in a *language* which is understood by the NC control system (including a computer where used), and in turn must be expressed on the control tape in the form of a suitable *code*. Programmes can be prepared manually for simple work (such as much point-to-point positioning), or prepared with the aid of a computer for more complex work (such as continuous path machining).

Depending upon the complexity of the NC machining operation, more than one type of programme may be required, these being as follows: –

a) *Part programme*. This is the programme (prepared by a person called the part programmer) which contains the complete information necessary to successfully machine the part on the NC machine tool. It must be written in a language which is compatible with the NC control system.

b) *Machine tool programme*. When the part programme is completed with the aid of a computer, a machine control tape will be produced as the final product of the computing operation. (see Fig 6.7). This is the tape which is fed to the machine tool control unit. This machine control tape is sometimes referred to as the Machine Tool Programme.

c) *Post-processor*. In a complex continuous path machining operation, say, a large computer may be used which produces a generalized solution defining the desired path of the cutter centre in the form of a complex series of ordinates. This mathematical information is coded onto a tape called a cutter location tape, and must be checked and put into a form which is acceptable to the control system of the specific NC machine being used for the part machining. A second computer is required for this job which has been prepared by means of a programme called a post-processor. This extra operation will then result in the production of a machine control tape (or machine tool programme).

Simply put, a post-processor (programme) is used to convert the general solution from a general-purpose computer into a specific solution encoded onto a tape which is acceptable to the control system of a specific machine tool.

The two common codes in use are BS 3635 and EIA as mentioned earlier in Section 6.5, and these at the moment appear to satisfy all NC requirements. The same is not true, unfortunately, of part-programme languages for NC applications. There is presently a proliferation of languages not one of which appears to be totally satisfactory. We will complete this section by outlining some of the more commonly used languages.

i) APT (Automatically Programmed Tools) is a universal part-programme language which can specify any component and its machining operations to be carried out on NC machines having up to five controlled axes. APT is a powerful and complex language which requires the use of a large computer, and is not convenient for simple applications. It uses the nomenclature of FORTRAN (Formulae Translation) which in turn is a universal computer language developed by IBM.

ii) ADAPT was the first of several alternative languages developed from APT. It is a simplified version of APT suitable for NC machines classified as 2CL.

(iii) PROFILEDATA is a 2CL language developed by Ferranti Ltd. in Great Britain as opposed to APT which was developed in U.S.A. It is used on control systems which use a magnetic tape as the input source.

iv) EXAPT (Extended subjects of APT) is an important language developed in Germany, which is made up of three parts:

EXAPT 1 for point-to-point applications;
EXAPT 2 for turning work;
EXAPT 3 for 2CL milling type work.

v) NELAPT is a programming language based on APT and developed by the National Engineering Laboratory (NEL).

Finally, it is interesting to note that languages used for continuous path machining applications make use of a 'pidgin English' type of vocabulary to give cutter motion instructions. Examples are:

FEDRAT = Feedrate
GOFWD = Go forward
GOLFT = Go left
TANTO = Tangential to

6.9 PLUG BOARD SYSTEM

So far in this chapter we have been concerned with numerical control systems, i.e., where the required ordinate dimensions and also the machining instructions are fed into the system using binary numbers on tape. In this section we will consider a plug board machine control system, although it is not a complete numerical control system, being rather a compromise between an automatic machine and a numerical control machine. However, a plug board system does give fully automatic control of a machine tool.

These systems are rapidly gaining in popularity because they are flexible and reliable, but are comparatively cheap when fitted as an 'attachment' to an existing machine tool.

If we consider the case of a capstan lathe, for example, the operator can be replaced by a plug board system, which can be programmed to suit the machining of a particular component. We then have an automatically operated capstan lathe which will be similar in operation to a turret automatic. The machine slides can be actuated by any suitable device, such as a hydraulic cylinder, or a pneumatic cylinder for example. These cylinders will be operated at the appropriate time according to the plug board programme, and hence the slides carrying the tools will move as required at the chosen speed through the required length of stroke. In other words, the machine receives a series of instructions in order to complete the component, and each instruction may involve one or more machine functions such as speed selection, feed rate, tool selection, tool pause, etc.

Fig 6.18 is a diagram of a plug board programming unit.

This unit is a very small one having the means for selecting three instructions, 1, 2 and 3 and three machine functions, A, B and C. In practice, plug boards are produced having the means of selecting 30 instructions, and 40 functions, say.

Reference to Fig 6.18 shows that with the three-way stepping switch set in the first position (as drawn) functions A, B and C are available

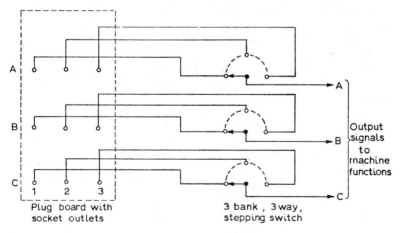

Fig 6.18 Plug Board Programming Unit.

for instruction No. 1. Now if plugs are pushed into sockets A and C say in column 1, then these circuits are activated, and hence functions A and C will operate on the machine during operation (or instruction) No. 1. When instruction No. 1 is completed, then the stepping switch automatically moves to the second position which makes all the functions, if required, available for instruction No. 2. If only B and C are required for this second operation, say, then plugs are pushed into these sockets only, and so on. The arrangement of plugs in all the socket outlets will be determined by the programme and will be set up before production commences. The programme will obviously be determined by the component and the necessary sequence of operations involving several or all of the machine functions.

A *bus bar* system is usually used for the plug board, two sets of bus bars being arranged at right angles to each other. One bus bar is required per horizontal row, or vertical column of plug sockets. Each bar is connected up to the appropriate stepping switch contact according to the arrangement shown at Fig 6.18. There will be one plug socket at each intersection of a horizontal and vertical bus bar.

Finally, let us consider the production of a simple component on a plug board operated capstan lathe to illustrate the principle of programming. The component is a simple length of bright free cutting, mild steel which has to be parted off at the appropriate length. With regard to the machine functions we will assume that the turret slide (carrying the bar stop) is operated by a pneumatic cylinder, as is also the cross slide (carrying the parting off tool). When they reach the stop after traversing, they operate a microswitch which then causes the stepping switch to move to its next position, hence continuing the sequence of operations. The collet chuck is opened and closed by an air cylinder operating its lever, and there are two spindle speeds available (high and low) which can be selected by switching a two-speed motor. A 'Home' function must be incorporated, such that when it is activated in the circuit, everything returns to its original position ready for the next cycle.

The sequence of instructions or operations can then be written as follows:

Instruction	*Function*
No. 1.	Select high spindle speed.
	Start spindle.
	Advance turret.
No. 2.	Open chuck (bar will move up to stop under action of weight).
No. 3.	Close chuck.
No. 4.	Advance cross slide.
No. 5.	Home.

(Note. When the slides reach the end of their stroke, and operate a pre-set microswitch, they will automatically be returned to their starting position as required. Hence, retracting a slide does not count as a machine function.)

This set of instructions can be written as a simple programme thus, where a vertical column represents an instruction, and a horizontal row represents a function.

		1	2	3	4	5
START SPINDLE	A	X	X	X	X	
SPINDLE SPEED (HIGH)	B	X	X	X	X	
SPINDLE SPEED (LOW)	C					
OPEN CHUCK	D		X			
CLOSE CHUCK	E			X		
ADVANCE TURRET SLIDE	F	X	X	X		
ADVANCE CROSS SLIDE	G				X	
HOME	H					X

Plugs will be placed in the sockets of the plug board to match this programme, a plug being placed where an *X* is shown in the programme.

An actual programme for a complex component might require a complicated pattern of plugs to be set up on a plug board having say 40 rows × 30 columns. If the component is required in many batches over a long period of time, then the programme can be permanently wired on to a board which can be stored when not in use. This of course is an expensive method. Alternatively, to make setting easier, the programme can be punched on to a card which fits over the plug board, leaving only the plug sockets exposed which are required for the programme. The card can then be filed for later use.

Two last refinements might be mentioned. One is that indicator lamps can be provided above each vertical column on the plug board to show at a glance which column on the programme is functioning at any time. The other is that electromechanical counters can be used to indicate how many components have been produced by the machine. They can be pre-set so that the machine will automatically produce a given batch quantity and will then switch off upon completion of the batch.

Exercises 6

1. Differentiate between a closed loop and an open loop control system for machine tools. Give an example of each.

2. Differentiate between a decimal numerical system and a binary numerical system. Which forms the basis of a numerical control system for machine tools and why?

3. Explain how numerical information can be stored upon (a) a punched tape, (b) a punched card, and (c) a magnetic tape. What is the advantage of each method?

4. (a) Convert the following decimal numbers into equivalent binary numbers: (i) 13, (ii) 28, (iii) 123, (iv) 572.

(b) Show the pattern of punched holes on a 4-track tape to represent the value 15·695 3.

5. Show by means of a block diagram the series of steps involved in carrying out continuous path machining by numerical control system.

6. With the aid of diagrams, describe a fully automatic point to point machining system for drilling components using (a) tape control, or (b) decade dial control.

7. Show, with the aid of sketches, how the position of the cutter relative to the workpiece can be accurately monitored on an automatically controlled machine tool. Describe two systems, one being digital, and the other being analogue in principle.

Comment upon the accuracy of each.

8. What are diffraction gratings? Describe, using sketches, how they can be used as a means of monitoring slide displacement upon a machine tool.

9. Show the principle of operation of (a) a recirculating ball leadscrew and nut, and (b) a hydrostatic linear bearing, and a hydrostatic rotating bearing.

Why are machine tool refinements such as these important to the successful automatic operation of machine tools?

10. An 18 mm diameter × 1·5 mm pitch threaded bolt is to be produced upon a plug board capstan lathe using hexagon mild steel bar. The bolt is 80 mm long. Draw up a programme for the production of the bolt, using the machine functions shown in the example at the end of Section 6.8. Assume the plug board is of any desired size.

Further Reading

1) Bryan G. T. *Control Systems for Technicians*. Hodder and Stoughton Ltd.

2) West J. C. *Servomechanisms*. Hodder and Stoughton Ltd.

3) *Machine Tool Automation by Electronic Control.* Yellow Back Series No 38, Machinery Publishing Co.

4) De Barr A. E. 'The development of Numerical Control'. *Journal of the Institute of Production Engineers*, Sept. 1971.

5) 'An Engineer's Management guide to computers'. *Journal of the Institute of Production Engineers*, Nov. 1974.

6) Martin S. J. *Numerical Control of Machine Tools.* Hodder and Stoughton Ltd.

7) Wightman E. J. 'Developments in Computer Control Systems for Machine Tools'. *Proc. Institute of Mechanical Engineering*, June 1974.

8) Ferguson N. C. 'A history of Numerically Controlled machine tools'. *Journal of the Institute of Mechanical Engineers*, Sept. 1978.

9) Charnley C. J. 'Some applications of CNC'. *Journal of the Institute of Production Engineers*, Jan. 1976.

10) 'Getting the most out of CNC'. *Journal of the Institute of Production Engineers*, Oct, 1977.

CHAPTER 7

Generation of Forms

7.1 FORMING v GENERATING

WHEN PRODUCING plain shapes on a workpiece using a machine tool, few difficulties arise. If a profile of unusual shape is required however, it can present a difficult technical problem. The complexity of the shape and the quantity of parts required will affect the choice of method used. In general a profile on a component can be *formed* or *generated*.

Forming
In this case a special form tool is required which bears the required profile upon its cutting edge. The shape produced on the workpiece will be the reverse image of the tool shape, as shown in Fig 7.1. This shows a form being produced upon a part being turned in a centre lathe.

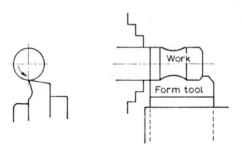

Fig 7.1　Forming upon a Lathe.

This diagram shows a simple, flat form tool, but on a production machine such as an automatic a circular form tool would be used. These have a much greater life, giving many more re-grinds, and are often easier to manufacture.

There are some important points to consider about form machining:
a) The accuracy of the work profile is entirely dependent upon the accuracy of the form tool profile.
b) The tool setting is important. If it is not set on the work centre, in the case of turning, then the profile will be incorrect.

c) Small form tools are often made without top rake. (Clearance, of course, is essential, as any tool will cut without rake but not without clearance.) If top rake is essential for efficiency of cutting, as shown in Fig. 7.1, then the profile transmitted on to the work will be slightly different than that ground upon the tool. Sometimes the difference does not matter on production work, but where it does, the tool must be *corrected* or modified in order to make the work profile correct.

d) Form tools do not cut too efficiently, and a long profile can cause chattering and poor finish. Cuts must be light. Form milling cutters will always have a tendency to chatter because of the inadequate clearance behind the cutting edge leading to rubbing between cutter and work. The reason for the inadequate clearance is that the cutter teeth are form relieved, as shown in Fig 7.2, in order to allow several re-grinds on the tooth face without destroying the tooth form. Form relieving produces a spiral tooth shape behind the cutting edge.

e) Complex form tools or cutters of any type are difficult to produce and are correspondingly expensive.

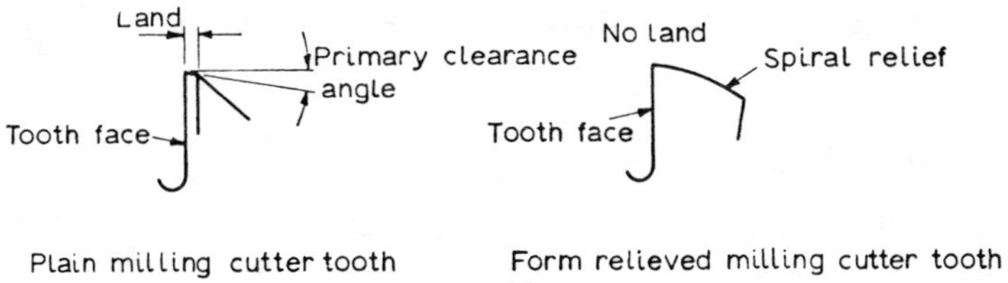

Fig 7.2   Comparison of Plain and Form Relieved Cutter Teeth.

Generating

In this case the tool shape has no effect upon the finished workpiece profile. The form of the profile is produced by the cutting edge of the tool being constrained to move through the required path, the required profile therefore being generated upon the work. This is the principle of the copying systems discussed in Chapter 4.

The hydraulic copying system shown at Fig 4.38 illustrates the principle of form generating. The following points are worth consideration:

a) In generating, the work profile is independent of the tool shape and the tool cutting angles.

b) It is still important that the tool is set correctly. Imagine the effect of the tool set well below centre when generating a long taper on a lathe using a taper turning attachment.

c) Much longer profiles can be generated than can be formed. Also the work finish will be better. It is generally easier to generate an internal profile than to form it.

d) Equipment which will generate profiles is usually expensive in the first instance, but will prove economic on production work because of the lower variable costs and the higher quality of work. Once templates or patterns are made, they can be used again for any number of batches.

Figure 4.37 shows an interesting example because it is a process which is part generating and part forming. The grinding wheel profile is generated, the wheel then being used to form the profile upon the work.

Much engineering work which is machined is often a mixture of generated and formed surfaces. The part shown in Fig 7.1 for example, has a formed shape looking on the side view, but has a generated shape (by virtue of the spindle rotation) looking on the end view. A screw thread cut with a single point tool on a centre lathe, has the thread shape or profile formed, and the helix of the screw generated. In this chapter we will consider the various methods of generating the important classes of formed work on a quantity production basis. We will restrict it to considering thread production and gear production. It is assumed the reader has a knowledge of the basic principles of thread design and gear design and is also familiar with the use of hand taps and dies, screwcutting on a centre lathe, and the form milling of single gears on a dividing head.

7.2 THREAD CHASING

This method is similar in conception to single point tool screw cutting on a centre lathe, but uses a chasing tool which is in effect several single point tools banked together in a single tool called a chaser. The method can be carried out at speed on a capstan lathe or turret lathe equipped with an interchangeable lead screw called a *leader,* which is driven from the feed box. Thread chasing is usually used on production for threads which are too large in diameter for a die head. It is not confined to external threads only, but can be used for internal threads above approx. 25 mm diameter. The principle is shown in Fig 7.3.

Reference to Fig 7.3 shows a tangential type chaser being used for cutting an external thread, and a circular chaser being used to cut an internal thread. The system is so arranged that the chaser traverses away from the headstock whilst machining, thus preventing the possibility of the chaser being accidentally allowed to run into the chuck while traversing. For RH threads the spindle must therefore be rotated in reverse.

The chaser is advanced radially into the work for each cut by means of the cross slide screw, several passes sometimes being needed to complete the thread. This cross feed movement is independent of the quick withdrawal mechanism which operates when the leader nut is withdrawn outwards.

A standard set of interchangeable leaders and mating nuts of varying pitches are provided, which with the small range of speeds available in a capstan lathe feed gear box ensures a wide range of threads can be cut. For example, say leaders having pitches of 6 mm, 5·5 mm, 5 mm, 4·5 mm, 4 mm, 3·5 mm and 3 mm are available.

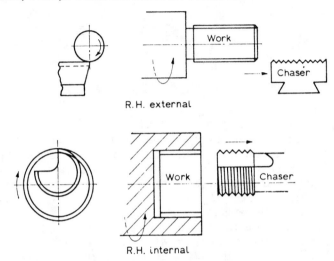

Fig 7.3 Thread Chasing.

With a three-speed gear box having ratios of 1 – 1, 1 – 2 and 1 – 4, the following pitch threads can be cut:

3 mm to 6 mm in steps of 0·5 mm

1·5 mm to 3 mm in steps of 0·25 mm

0·75 mm to 1·5 mm in steps of 0·125 mm

As would be expected, chasing lends itself better to non-ferrous materials rather than ferrous. Multi-start threads can be chased without any indexing of the workpiece being necessary. Taper threads can be generated by chasing, if the chasing attachment is used in conjunction with a taper turning attachment.

7.3 DIE HEADS

Self opening die heads or die boxes are used extensively for the high production of external threads on capstan and turret lathes and all types of automatics. The heads are containers of varying sizes which hold dies or chasers, each head being suitable for a given range of sizes. Heads are available for cutting threads from 6·35 mm diameter to 114 mm diameter. Chasers are available for any thread form, and using the correct type of chaser, any type of material can be screwed including some of the plastics, such as bakelite.

There are three types of die heads, the difference depending upon the types of dies used. They are those having (a) *radial dies,* (b) *tangential dies,* and (c) *circular dies.* Each type is illustrated in Fig 7.4.

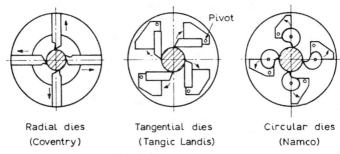

Radial dies   Tangential dies   Circular dies

(Coventry)   (Tangic Landis)   (Namco)

Fig 7.4 Die Heads.

The dies open automatically when the required length of thread is cut. When the turret slide movement is arrested by a stop, the front part of the head continues to move forward by a small amount until the dies spring outwards, away from the work under the action of a scroll or cam. The Coventry die head having radial dies is the most commonly used general purpose die head.

The other types are used for more special applications, particularly for threading difficult materials. Provision is provided on all types for taking roughing or finishing cuts by moving a detent pin to the appropriate position. Regrinding of dies must be done in the special fixtures provided in order to maintain the correct throat and lead angles.

When used upon a capstan lathe, the operator need only keep the die head up to the work by applying slight pressure to the turret slide capstan handle. The die head is self guiding due to the action of the dies which cut at the throat only, the rest of the die thread acting as a guide nut. Hence, the die head will screw itself along the work until the dies trip open. The work spindle does not have to be reversed in order to screw the die head off again, and production rates are increased as a result. The dies can be closed by the operator after each screwing operation, by pushing a handle which partially rotates the front portion of the head. If threads of very high quality are required, the die head can be used with the chasing attachment by engaging the leader screw nut of appropriate pitch during threading.

When used on an automatic, the die head feed motion is controlled by the cam rise which is designed accordingly. After threading the dies are automatically closed by arranging for the closing handle to strike a rod on the return stroke.

Die heads for internal threads are in fact, collapsible taps. These are similar in principle to a Coventry die head in that they have radial chasers. These withdraw, or collapse inwards when a hardened steel ring around the tap strikes the end face of the work. Hence, again there is no necessity to reverse the spindle in order to withdraw the tap, and much time is saved. They are suitable for threading holes above 25 mm diameter.

7.4 THREAD ROLLING

Threads can be rolled to shape by cold working in dies, and no further treatment is necessary, the thread then being complete. The process is most often applied to precision threads, and splines are often rolled to shape on production by this method. The process consists of cold flowing of the material through dies under the application of compressive forces. The material is progressively forced down to the root diameter of the thread, the excess material flowing upwards within the die space to form the upper portion of the thread. The thread blank is made to the mean diameter of the thread, giving a saving of material. The choice of material is important, because the amount of cold working it can tolerate will depend upon its ductility and rate of work hardening. Generally speaking, metals having not less than 15% elongation and not more than 300 BHN hardness will cold work satisfactorily.

As with all cold working of metals, thread rolling produces a bonus in terms of material properties. The surface hardness, strength and fatigue resistance of the thread will all be improved. The grain flow, which follows the surface contour, will be ideal, and the surface finish will be good (approximately $0.2\,\mu$m R_a). The improvement in properties means that a cheaper material can be used for the blanks in the first instance, hence reducing variable costs. The tools used are generally expensive, but die life is long and production of accurate threads can be maintained at high speeds over a long period.

Thread rolling can be accomplished using either *flat dies*, or *circular dies* contained in a die head.

Flat Dies

These are used in a special production machine which can be fully automatic, requiring only the hopper filling and capable of high speed production. The principle of operation is shown at Fig 7.5.

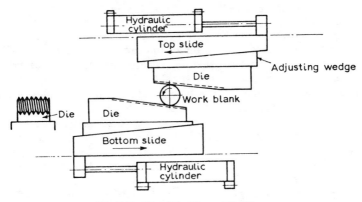

Fig 7.5 Flat Die Thread Rolling.

This diagram shows both dies moving under the action of hydraulic cylinders, although on some designs only one die traverses. The thread is formed complete after one pass of the blank between the dies. Each die is grooved with the thread profile, the thread grooves being inclined at the thread helix angle to avoid interference during rolling. The blanks can be fed automatically (or manually) from a hopper between the dies, and the die length is such that the blank rotates about four times during one pass. The dies are made from high carbon, chromium steel, and have a phenomenal life (up to 500 000 parts) due to being subjected to little friction during rolling. Rolling speeds are between 0·50 m/s and 1·25 m/s depending upon the material properties.

Circular Dies

These are contained in a die head similar to the Coventry die head, and the rolling die heads can be used upon capstan and turret lathes, or automatics. They automatically open upon completion of the thread and need not be unscrewed off the work. They are not as fast in operation as flat dies, (which are almost instantaneous) but are not limited in the length of thread they can traverse. Flat dies can only roll threads which are not longer than the width of the die. A diagram of a rolling die head is shown at Fig 7.6.

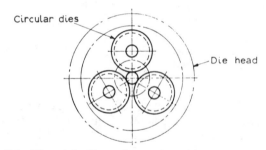

Fig 7.6 Thread Rolling Die Head having Circular Dies.

The dies shown in Fig 7.6 have no cutting edges of course, but are highly polished thread rolls mounted on large, friction free spindles. The thread grooves in the rolls are not annular, but are true helical grooves. The die boxes are not suitable for large work, as the rolls must be larger in diameter than the work, say in the ratio 3 − 1. As the roll is larger in diameter than the work, the helix angle of the roll thread groove will be different than that of the work thread groove, although the pitch and thread form are the same. Hence, interference could occur on the thread flank during rolling. To avoid this, the roll thread is made multi-start in the same ratio as the roll to work diameter ratio, therefore keeping the helix angles equal. For example, if the work is 12·5 mm diameter, and the rolls are 37·5 mm diameter, then the thread rolls will have a three start thread.

In action on the capstan lathe, the die head threads itself along the work, due to the helical lead of the dies, and does not need the traversing action of a leadscrew and nut. An oil lubricant is used. Some high production machines use circular dies which are independently driven, rather than traversing flat dies.

7.5 THREAD MILLING

This is a fast production method of cutting threads, usually of too large a diameter for die heads. As the milling cutter is held on a stub arbor, the length of thread which can be cut is limited, although threads running up to a shoulder present no difficulty. The cutter used is called a *hob* and has annular thread grooves and form relieved teeth. Large threads, such as a worm, are milled using a form cutter, but this process will be dealt with in Section 7.10.

A diagram of the process is shown at Fig 7.7.

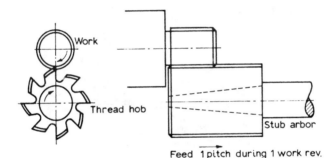

Fig 7.7 Thread Milling.

The hob rotates at the correct cutting speed for the operation (say 0·6 m/s), and the work slowly rotates at the correct feeding speed. As the work rotates the hob also feeds endways under the action of a master leadscrew, such that as the work rotates one revolution, the hob will traverse one thread pitch. As the hob is initially fed in to full thread depth over the full length of thread, the process is complete after one revolution of the workpiece.

Although the hob has annular grooves, it is not set over at the thread helix angle, because interference is negligible at the large diameters and fine pitches used in this process. A RH or LH thread can be produced by varying the direction of hob feed and direction of rotation of work. Internal threads can be milled, although interference will then present a problem due to the longer length of engagement between hob and thread.

7.6 THREAD GRINDING

This method is similar in principle to thread milling in that a grinding
wheel having annular thread grooves formed around its periphery cuts
the thread as wheel and work rotate. In both cases the process is one
of forming and generating. Modern grinding wheel technology has
enabled wheels to be produced which will maintain the thread form on
the wheel, and the process therefore owes something to the skill
of the grinding wheel manufacturers. A vitrified bond is generally
used with a fine grit up to a value of about 600. A special grinding
machine is necessary having a master leadscrew and gears and the means
of holding the work.

Two variations of the process are used, viz., (a) *traverse grinding,* and
(b) *plunge cut grinding.* In both cases the grinding wheel has annular
thread grooves around its periphery and can be set over at the thread
helix angle if necessary to avoid interference. The wheel rotates at the
correct speed for the operation (say 30 m/s), while the work rotates
slowly as the metal is removed.

a) Traverse Grinding
See Fig 7.8.

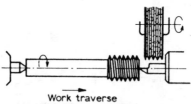

Work traverse

Fig 7.8   Traverse Thread Grinding.

The wheel is positioned at full thread depth, then the work is traversed
past the wheel. The work table traverse is controlled by a master leadscrew. Change gears are used to suit the thread pitch (as in screwcutting
on centre lathe). The first thread form on the wheel removes the majority
of metal and is therefore subjected to the most wear. The following
threads effect the finishing. A single ribbed wheel may be used for large
threads or special threads.

b) Plunge Cut Grinding
See Fig 7.9.

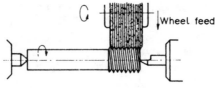

Wheel feed

Fig 7.9   Plunge Cut Thread Grinding.

The wheel is plunged into work to full thread depth. The workpiece then makes one revolution while the work traverses one pitch. This gives uniform wheel wear but is used only for short thread lengths.

Thread grinding can be used upon soft or hardened work and cuts threads from the solid. Hence the problem of thread distortion after hardening is not present.

Wheel Forming

An accurate thread profile must be produced upon the wheel face (as in all form grinding) and different methods are used by different manufacturers. Two basic methods are (i) *crushing* or (ii) *diamond dressing,* using some suitable means to guide the diamond such as a *pantograph* (as illustrated in Fig 4.37).

(i) *Crushing.* This makes use of a crushing roller of hardened steel which has the required thread form accurately produced around its periphery and is ground and lapped. The roller is fed into the wheel face under pressure using a great volume of lubricant, the thread profile being crushed into the wheel face, while the wheel rotates slowly.

The principle is illustrated in Fig 7.10.

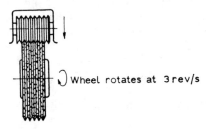

Fig 7.10   Grinding Wheel Crushing.

The crushing unit may be mounted on the wheel head above the wheel as shown, or on some machines the attachment is mounted upon the table. Care has to be taken otherwise excessive loads may be placed upon the wheel bearings. Crushing is a forming process.

(ii) *Diamond Dressing.* Different systems are used for generating the form upon the wheel face such as the pantograph described in Section 4.11. Messrs Coventry Gauge and Tool Co. Ltd. use an interesting patent method upon their 'Matrix' machines. This is illustrated in Fig 7.11.

As the leadscrew is driven from the machine, the cam rotates upon the spindle which is coupled to the leadscrew through change wheels. The cam imparts the longitudinal movement to the diamond. A suitable combination of cam lift and change wheels will give the desired thread profile upon the wheel face, such that as the machine slide traverses the diamond past the wheel, the leadscrew rotation will cause the diamond (via the cam) to trace out the thread profile.

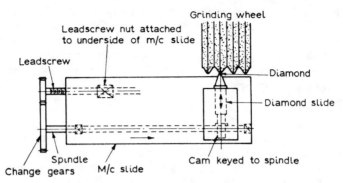

Fig 7.11 Diamond Dressing Attachment.

7.7 GEAR PLANING

Gears are machined in quantity by processes which generate the tooth profile. There are a variety of machining processes available, each of which has some particular advantage over the others. The gears we shall consider are: spur gears (external and internal), helical gears (single and double), racks, bevel gears (straight and spiral), and worms and worm wheels. From a production point of view, spiral gears may be considered as helical gears. Note that two involute gears of the same pitch will mesh with each other, and each gear will also mesh with an involute rack of the same pitch.

Gears can be planed using the Sunderland process or the Maag process, which are identical in principle but different in the machine configuration and detail. The planing process makes use of a cutter of true involute rack shape, having rake and clearance, which reciprocates across the face of the blank. Figure 7.12 shows a diagram of a rack cutter. The cutter and blank are geared together so that the blank rotates in the correct relationship to the longitudinal movement of the cutter, i.e., they roll together as a rack and pinion. This is shown diagrammatically at Fig 7.13.

It will be seen from Figure 7.13 that the tooth height of the rack

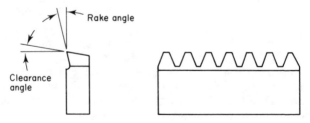

Fig 7.12 Rack cutter.

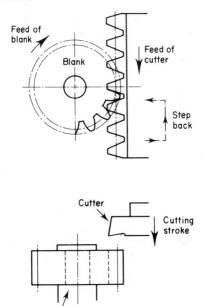

Fig 7.13 Gear planing.

cutter is twice the tooth dedendum, in order that the cutter touches the root circle and clears the top circle of the gear being cut.

The machining cycle is:

a) Cutter fed into full tooth depth with cutter reciprocating and blank stationary.

b) Blank rotates and cutter feeds longitudinally. An involute shape is generated on the gear teeth flanks by the involute rack cutter.

c) After one or two teeth are completed the blank and cutter stop feeding. The cutter is withdrawn and indexed back to its starting position.

d) Cutter fed to depth and cycle repeated.

The indexing, or 'stepping back' operation described at (c) above enables a short rack cutter of a practical length to be used. For faster production, two cutters can be fitted cutting on alternate strokes, thus machining two blanks at one setting. The idle return stroke is eliminated. Gear planing is used for the generation of external spur gears, and where the cutter slide is provided with angular adjustment external helical gears may be cut. If a rack cutter of standard proportions is used for cutting helical gears, then the normal pitch of the helical gear will be slightly incorrect.

The process is ideally suited for cutting large, double helical (herringbone) gears which do not have a clearance recess between the two sets of teeth. A special version of the planing machine is used in this case having two cutter slides inclined at the gear tooth helix angle.

In gear generation by planing, the rack cutter controls the pitch and pressure angle, and the machine gearing controls the number of teeth produced on the gear.

7.8 GEAR SHAPING

Gears can be shaped using the Fellows process, which is the most versatile of all the gear cutting processes. The shaping process makes use of a hardened pinion as a cutter, ground with top rake and clearance. Figure 7.14 shows a diagram of a pinion cutter. The cutter reciprocates across the blank face, and can cut either on the upstroke or downstroke. A relieving mechanism enables the cutter to clear the work on the return (non-cutting) stroke. The cutter and blank are geared together so that they rotate in the correct relationship to each other to suit their respective numbers of teeth, i.e., they rotate together as two gears in mesh.

The principle of operation of the process is shown at Fig 7.15

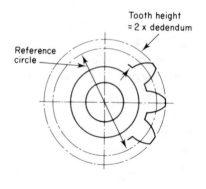

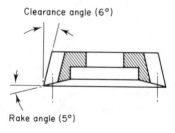

Fig 7.14   Pinion cutter.

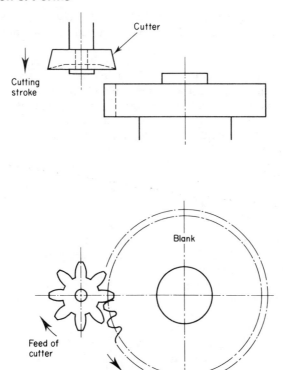

Fig 7.15 Gear shaping.

The machining cycle is:
a) Cutter fed into full depth, with cutter reciprocating and blank stationary. (Blank could be rotating if desired.)
b) Cutter and blank slowly rotate until all the teeth are generated upon the gear blank.

As with most gear cutting, the blank is usually roughed to say three-quarter depth, followed by a finishing cut. The process can be used for cutting external and internal spur gears, and racks if used with a rack cutting attachment. All these gears can be generated in helical form, but a helical cutter of the correct hand (opposite to the required gear hand) must be used in conjunction with a helical cutter guide of the same angle and hand as the cutter. Usually standard helix angles of 15° and 23° are available, and the process is not recommended for helical (and spiral) gears having a helix angle greater than 35°. Referring to Fig 7.15 when cutting helical gears the cutter is constrained to follow a helical path, instead of the vertical, straight line path shown in this diagram. As a special cutter must be used, it is manufactured with the correct normal pitch to suit the helical gear.

This fundamental generating process of work and cutter rotating at constant angular velocities about fixed centres is a form of conjugate machining and gives great versatility. In addition to the range of gears already mentioned, many special shapes such as polygons or cam profiles can be cut. For example a four leaf, clover pattern cutter will generate a square upon the workpiece, if cutter and work rotate at the same velocity.

7.9 HOBBING

This is the fastest of the gear generating processes as it is a continuous cutting process. The reciprocating or indexing movements in the other gear cutting machines are absent from a hobbing machine. This is one of the most fascinating generating processes, and it is difficult to visualize it working until one actually sees the process being carried out.

It may be helpful to first imagine a disc being generated using a slab or roller milling cutter as shown in Fig 7.16.

The milling cutter is rotated at the correct cutting speed to suit the disc material, and is slowly traversed across the face of the disc at the required depth of cut. The disc rotates and will therefore be machined to a reduced diameter. In Fig 7.16 the process is one of up-cut milling, but could equally well be down-cut. Note that there is a distinct relationship

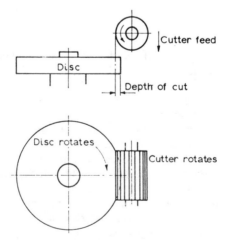

Fig 7.16   Generating a Disc by Milling

between the cutter feed, and speed of disc rotation. If the milling cutter is fed too quickly across the disc face, then a helical groove will be milled on the disc face.

If we wish to generate gear teeth upon the disc, we must replace the milling cutter with a form relieved milling cutter called a hob. This is in effect a worm having a thread of involute tooth rack shape, gashed with form relieved teeth. Figure 7.17 illustrates a gear cutting hob in half-section with the rack profile of the worm thread being shown.

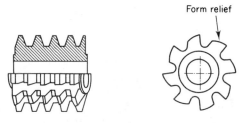

Fig 7.17 Gear hob.

The hob is rotated at a suitable cutting speed and fed across the blank face. The hob and blank are geared together so that they rotate in the correct relationship to each other, i.e., they rotate as a worm and worm wheel in mesh. In effect, the worm as it rotates presents an endless set of rack teeth to the blank.

The principle of hobbing is shown at Fig 7.18.

The machining cycle is:

a) Hob and blank rotate.

b) Hob fed into full depth.

c) Hob feeds across blank face and generates involute teeth.

As with all milling-type operations, there must be sufficient clearance on the side of the blank to allow the hob to run out clear of the work. Reference to Fig 7.18 shows one or two important points. The hob must be tilted through its helix angle α when machining straight spur gears, in order to bring the hob teeth into line with the cut gear teeth and avoid interference. Therefore the rack form of the teeth must lie along a line at angle α to the hob axis as shown by section *A.A.* The teeth gashes are usually machined in the hob normal to the tooth helix as shown. Multi-start hobs are used for fast roughing upon production, because a three start hob for example, will cut three gear teeth for one revolution of the hob. The more accurate single start hobs are used for finishing, and one hob revolution generates one gear tooth. Hobbing can give the highest

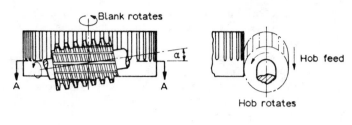

Part section A-A
(True involute rack shape)

Fig 7.18   Gear Hobbing.

rate of gear production, and the fastest machines are fitted with fully automatic loading and ejecting equipment.

The hobbing process is versatile and helical and spiral gears can be produced upon a fully universal hobbing machine. A standard hob can be used for cutting helical as well as straight gears, but the hob must be tilted through an additional angle equal to the helical gear helix angle, in order to avoid interference (i.e., if both hob and gear have RH helices). In this case, as when using a standard cutter in planing, the helical gear normal pitch will be slightly incorrect. Also, it is necessary to use differential change gears when machining helical gears in order to give an auxiliary rotation to the blank. The hob will still feed down a straight, vertical line as shown in Fig 7.18, but the gear teeth are required to follow a helical line. Hence, the blank must continually accelerate as it were for each increment of hob feed.

Double helical gears can be produced by hobbing if a clearance groove is left between each set of teeth, in order to provide run-out for the hob.

Hobbing is the ideal process for generating worm wheels. The hob is made to correspond to the mating worm which is to be used with the worm wheel in service. The hob is manufactured to the top limit of diameter to allow for the hob teeth being re-ground. (Note that the teeth are re-ground on their front face, but the hob diameter will decrease after each re-grind due to the form relieved shape of the teeth.) Large worm wheels are usually roughed out using a fly cutter and then finished hobbed. The hobbing process for worm wheels is identical in principle to that used for producing gears, but the hob can be fed to full tooth depth either radially or tangentially.

Spline Hobbing
If the hob tooth profile is made other than involute rack shape, then shapes other than involute can be generated by hobbing. Serrated or

splined shafts, for example, can be hobbed and frequently are when required in large quantities. External splines can be more accurately and quickly generated than formed. The required hob tooth profile to produce a standard spline can be either plotted by graphical means on the drawing board, or the points on the developed tooth curve can be calculated.

The tooth profile and the resultant spline shape are shown sketched in Fig 7.19.

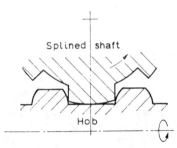

Fig 7.19 Spline Hobbing.

7.10 WORM MILLING

Worms are usually produced by milling the worm thread in the blank on a special worm milling machine using a form relieved milling cutter. The operation is as screw cutting on a centre lathe, but a cutter is used instead of a tool.

The worm blank is held on a mandrel between centres (for worms with central hole) or in a collet chuck and supporting centre for solid worms. It is traversed past a rotating cutter which is carried in an adjustable head. The cutter is set over to the helix angle of the worm being cut to avoid interference. Change wheels are used to give the correct worm lead, and indexing is used for multi-start threaded worms. The cutter form must correspond to the worm tooth space in the normal plane. Fig 7.20 shows the process.

The milling process described is commonly used for all general worm production, but hobs can be used for hobbing worms although this method is only recommended for worms having a large helix angle. It is much faster than the milling process however. Hob teeth are made with the same pressure angle and normal pitch as the worm.

Worms are often hardened and then ground, the grinding operation being identical to the milling process in principle but using a form grinding wheel. 0·15 mm to 0·25 mm is left as a grinding allowance.

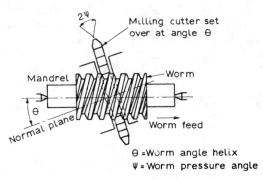

θ =Worm angle helix
Ψ = Worm pressure angle

Fig 7.20   Worm Milling.

Gear Finishing

After spur or helical gears have been machined from the blank stage they may require a finishing process to be carried out on the teeth if high class, quiet running gears are required. There are two main methods of approach:

1.	2.
(a) Machine gear teeth (plane, shape or hob).	(a) Machine gear teeth (plane, shape or hob).
(b) Shave gear teeth.	(b) Harden if required.
(c) Harden if required.	(c) Grind gear teeth.
(d) Lap using master gear plus abrasive lapping compound.	
	(a) is a primary machining process.
(a) is a primary machining process.	(c) is a finishing process.
(b) and (d) are finishing processes.	

We have considered the primary machining processes, and in Sections 7.11 and 7.12 we will consider the main finishing processes.

7.11 GEAR SHAVING

A high speed rotating shaving cutter is pressed into mesh with a free running gear to be shaved and the cutter is traversed across the face of the gear. The cutter is copiously fed with cutting oil and the minute shavings removed by the cutter are washed clear. The process is simple and production machines have an automatic cycle (see Fig 7.21). The cutter details are shown at Fig 7.22.

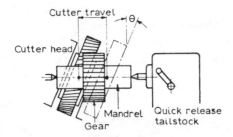

Fig 7.21 Shaving Gears by the Axial Traverse Method.

θ = Angle of inclination of cutter.
= Helix angle of cutter for shaving spur gears
OR
= Helix angle of cutter plus gear helix angle for helical gears.

The cutter feed may be *axial* (as shown in Fig 7.21), *oblique* or *tangential* to the machined gear. The first is simplest, the second gives most even cutter wear, and the latter is most suitable for shouldered gears where run-out is limited.

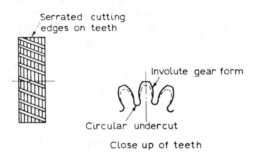

Fig 7.22 Close up View of Shaving Cutter Teeth.

The circular undercut between the teeth shown in Fig 7.22 provides clearance for the shavings 'filed' off the gear tooth flanks.

7.12 GEAR GRINDING

As outlined previously, hardened gears may be lapped if required, but many high class, hardened gears are ground. The gear grinding process may be a forming operation in principle, as shown in Fig 4.37, or a generating operation. As this chapter is concerned with generation of shapes by machining, we will look at the generating process in detail.

The Maag process is the most commonly used. It makes use of a similar principle to that used for rack planing of gears, but the gear rolls past the grinding wheels which have flat sides representing the basic

involute rack. Also the gear is traversed past the wheels. The principle
of operation is shown at Fig 7.23.

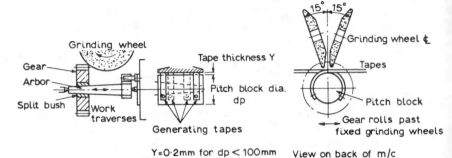

Y=0·2mm for dp < 100mm   View on back of m/c
Y=0·3mm for dp > 100mm

Fig 7.23   Gear Grinding using the Maag process.

The rolling motion of the gear blank (and hence the arbor and
driving spindle) is controlled by the pitch block which is pulled along
its guides by steel tapes. Theoretically the saucer grinding wheels
should be inclined at the gear pressure angle, and the pitch block
diameter (less tape thickness) would be equal to the gear PCD, but in
practice Maag wheels are fixed at an inclination of 15° and the pitch
block diameter is modified accordingjy, thus:

$$dp = \left(\frac{d \, Cos \, \psi}{Cos \, \alpha}\right) - Y \text{ where}$$

ψ = gear pressure angle
α = grinding wheel inclination
d = gear PCD
dp = pitch block diameter
Y = tape thickness.

Example. In a Maag gear generating process, $\alpha = 15°$, $Y = 0·2$ mm, ψ
$= 14\frac{1}{2}°$, $d = 63·5$ mm, calculate the pitch block diameter dp.

Solution

$$dp = \frac{63·5 \times Cos \, 14\frac{1}{2}°}{Cos \, 15°} - 0·2 = 63·63 - 0·2$$

$$= 63·43 \text{ mm.}$$

The gear is slowly traversed past the wheels (meanwhile rolling and
generating the tooth form) until it is clear of the wheel. The gear is
indexed one tooth pitch to the next tooth space, then traversed back
and the next tooth form generated on the reverse feed. Dividing plates
are used as for the form grinding process. On the Maag process grinding
wheels are dressed during grinding.

This method can be used for grinding helical gears by swinging the
wheel head through the helix angle of the gear. This inclination is taken
into account when calculating the diameter of required pitch block.
Otherwise the helical gear grinding process is the same as for spur

gears. Generating processes are available similar to the Maag process, but make use of a rack and pinion mechanism instead of the pitch block and tapes for rolling the gear past the grinding wheels.

7.13   STRAIGHT BEVEL GEAR MANUFACTURE

The teeth of bevel gears constantly change in form from the large to the small end, the teeth flanks being involute as are spur gear teeth. Therefore it is impossible to obtain a theoretically correct tooth form by machining out the tooth space using a rotary type milling cutter having fixed curves, i.e., bevel gear teeth cannot be formed but must be generated. The process consists in practice of (a) a roughing operation which removes most of the metal between the teeth and (b) a finishing operation to give the correct involute form.

a) *Roughing.* All the teeth are roughed out at one setting, each tooth being indexed around in turn. This is a forming operation which uses a rotary milling cutter with involute rack tooth form The cutter must be made to the tooth space involute shape at the small end of the gear leaving the tooth requiring most finishing at the large end. The roughing operation may be done on a milling machine with dividing head or indexing fixture, or on a special roughing machine which uses large diameter inserted tooth cutters. The operation is shown at Fig 7.24.

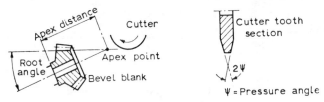

Fig 7.24   Straight Bevel Gear Roughing operation.

b) *Finishing.* This must be a generating process to give the correct involute tooth profile and may be carried out on a Bilgram or Gleason bevel gear generator. The process is similar in principle to an involute

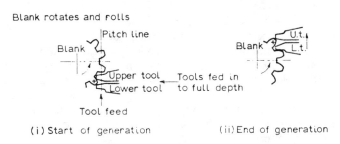

Fig 7.25   Straight Bevel Gear Finishing operation.

rack generating spur gear teeth by planing, i.e., two rack shape cutters plane the tooth flanks while the blank rolls around and rotates (see Fig 7.25). An enlarged view of the cutters used is shown at Fig 7.26.

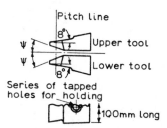

Fig 7.26   Enlarged view of Cutters used for Bevel Gear Finishing operation.

The machining cycle is as follows:
a) Tools fed in to full depth whilst reciprocating across blank face. [See Fig 7.25 (i).]
b) The roughed blank starts to rotate and roll, as the reciprocating tools feed along the pitch line.
c) One tooth is completely generated at the end of the feed stroke. [See Fig 7.25(ii).]
d) Tools are fed out, and indexed back to the start of the feed position.
e) Cycle repeated for each gear tooth.

High class straight bevel gears are finally lapped using a master gear and lapping compound.

7.14 SPIRAL BEVEL GEAR MANUFACTURE

These gears, and hypoid gears (a special version of spiral bevel gearing having offset gear axes), are used a great deal in the engineering industry, particularly in transmission units in motor cars. They may be required to be produced accurately in large quantities, and considerable ingenuity has been shown in the design of the machines used for generating spiral bevel gears.

These gears bear the same relationship to straight bevels that helical gears bear to spur gears. The Gleason spiral bevel gear generator may be used for their manufacture, but the teeth produced lie on the arc of a circle, instead of the theoretically correct part of a spiral curve. However, bevel gears produced by the same method of manufacture will mesh efficiently together. Each tooth space is generated separately while the blank rolls, the gear blank being indexed after the cutter is withdrawn from each tooth space and returned to its original position. Therefore one tooth is generated at a time.

The Gleason process uses a circular cutter with inserted teeth of the face mill type. The principle is the same as for straight bevels, i.e., the cutter flanks are of straight involute rack form, but in this case a cutter is used for both roughing and finishing. The gear blank and rotary cutter roll together exactly like two spiral bevel gears in mesh. See Fig 7.27 which is a sketch of the operation.

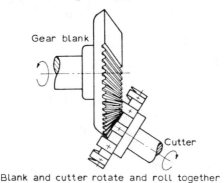

Gear blank

Cutter

Blank and cutter rotate and roll together

Fig 7.27   Spiral Bevel Gear Cutting.

Exercises 7

1. State whether the following machining operations are forming or generating, or a mixture of each: (a) milling a keyway using an end milling cutter, (b) drilling a hole using a twist drill, (c) finish shaping a flat surface using a broad tool, (d) face milling a flat surface.

2. With the aid of sketches show how a 40 mm diameter, LH, internal thread may be chased upon a capstan lathe. Show the direction of rotation of the spindle, the position and feed direction of the chasing tool. Sketch the chaser.

3. Contrast and compare the following three methods of producing external threads on components in quantity: (a) Coventry die head, (b) chasing, (c) rolling. State the advantages and disadvantages of each method.

4. Using diagrams to illustrate your answer, show the principle of operation of thread milling for (i) fine pitch threads, and (ii) coarse pitch threads such as worms.

5. Discuss and compare the different methods of machining plain spur gears (internal and external), helical gears (internal and external) and straight and helical racks.

6. Worm and worm wheel sets are required in large quantities. Describe suitable manufacturing methods and discuss the problem of interchangeability of worms and wheels.

7. The processes of shaving or grinding can be used to finish plain or helical gears. Describe and compare each process.

8. Using diagrams, show the two fundamental generating processes for producing (a) straight bevel gears, and (b) spiral bevel gears.

Further Reading

1) Houghton A. *Metal cutting tools.*

2) Merritt H. E. *Gears.* Sir Isaac Pitman, London.

3) Chapman W. A. J. *Workshop Technology. Part* 3. Edward Arnold Ltd.

4) Arnold J. 'Spline and Thread Rolling.' *Journal of the Institute of Production Engineers.*

5) Rochat F. 'Some Recent Developments in Spiral Bevel Gears.' *Journal of Institute of Production Engineers.*

6) Whiteley F. 'Gear Manufacture.' Published in six booklets by D.A.T.A.

7) 'The Book of the Coventry Die Head.' Alfred Herbert Ltd., Coventry.

8) 'Thread rolling – a neglected art.' *Journal of the Institute of Production Engineers.* March 1979.

CHAPTER 8

Production of Fine Machined Surfaces

8.1 THE GRINDING PROCESS

THE GRINDING process is an abrasive one, leading to an improvement in the linear and geometric accuracy of a component ($\pm$ 0·025 mm) and an improvement in the surface finish (0·1 μm R_a). Hardened or soft, ferrous or non-ferrous metals can be ground, and in addition non-metallic materials such as plastics, glass, rubber, etc., can be ground.

On production lines, work can be rough or finish ground efficiently at high speeds. Plain or flat surfaces are processed upon a surface grinder which may have a horizontal or vertical spindle combined with a reciprocating or rotary table. Internal or external cylindrical surfaces are processed upon either centreless type machines (described in Sections 4.9 and 4.10) or the more orthodox, but slower centre type external cylindrical grinder, or the internal grinder. Other special purpose production grinding machines are used for processing gears (Sections 4.11 and 7.12), threads (Section 7.6), splines, cam shafts, crankpins, etc. The grinding cycle may be automatic, and the whole process (if quantities justify it) may be fully automatic, only requiring feeding with components.

The quality (degree of excellence) of the work will depend upon the correct wheel being selected, balanced and trued, to suit the operation, and also upon the design of the machine. Production grinding machines must be robust in order to be vibration free, and the wheel spindle must run precisely true and central in its bearings. Special wheelhead bearings, often of the segmental type, have been designed over the years to achieve this. The emphasis today is upon air bearings, both for supporting wheel spindles and slides. Internal grinding spindles running in air bearings can rotate at a speed of up to 1 000 rev/s.

We will look now at the basic principles of the grinding process, and firstly examine the nature of the grinding wheel.

BS 4481: Part 1: 1969 recommends a certain standard system for identifying grinding wheels, to which we will refer.

1 Nature of abrasive. The abrasive grains in the wheel actually remove cuttings from the work. The main abrasives used are *silicon carbide* (C), which is most efficient when used for grinding soft materials, and *aluminium oxide* (A), which is most efficient when used for grinding hard materials.

2 Grain size. After crushing, the abrasive grains are graded into size by sifting through various size screens. Grain sizes range from 8 (very coarse) to 600 (very fine).

3 Grade. The strength of the bond is called the *grade* and is denoted by a capital letter. These symbols range from A (very soft) to Z (very hard), or more usually from E to S. They indicate the measure of resistance offered by the combined strength of the bond and abrasive grains to the stresses set up during grinding.

4 Structure. The proportion of bond to grains in a wheel can vary, and hence alter the *structure* of the wheel. The proportion of bond to total wheel volume can vary from 10% to 30%, giving either a very close abrasive spacing, or open abrasive spacing respectively. The purpose of structure is to provide chip clearance, the more open wheel structure providing more clearance. This prevents the cut chips from clogging or 'loading' the wheel face, and hence increases the efficiency of cutting. Wheel structure is denoted by a number ranging from 0 (very dense) to 14 (very open). However, the use or otherwise of a symbol to denote *structure* of the grinding wheel is optional.

5 Nature of bond. The *bond* is the tool post if you like, which holds the abrasive grains (cutting tools) in place. Four main types of bonds are used in modern grinding wheels, and these are listed below:

  i) Vitrified (V)—Used for over 75% of all grinding wheels made. All purpose bond giving high stock removal rate.

 ii) Resinoid (B)—A tough bond, used for high speed operations such as cutting off or fettling. Gives a very rapid stock removal rate.

iii) Rubber (R)—Used for centreless grinding control wheels, and for very thin cutting off wheels ('Elastic' wheels < 0·80 mm thick).

 iv) Shellac (E)—Used where a high degree of surface finish is required, and for light duty work. Occasionally used for slitting wheels (> 0·80 mm thick).

Other less common bonds used are silicate (S), magnesia (Mg) and reinforced resinoid (BF).

BS 4481 recommends that a grinding wheel be marked with the various symbols shown in the given order, so that the particular characteristic of any wheel can easily be identified, for example:

C 30 Q 8 B

means a wheel having a silicon carbide abrasive of 30 grain size, Q grade, 8 structure and of resinoid bond. Such a wheel might be suitable, say, for cutting-off or fettling cast iron components.

BS 4481 allows that manufacturers might also wish to add one or two further symbols to the marking system in order to identify other minor characteristics.

Wheel shape. Many standard wheel shapes are available ranging from plain discs and cylinders, straight and taper cups, to dish and depressed centre wheels. Most production grinding is carried out using plain disc wheels.

Wheel selection. When selecting a particular grinding wheel for an operation, there are many factors which should be considered. These can be grouped under constant factors or variable factors.

Constant Factors

a) Type of material being ground. (This is covered by the very important general principle of grinding, viz., 'when grinding a soft material use a hard grade of bond, and vice versa'.)

b) Type of grinding process, whether roughing or finishing.

c) Area of contact between work and wheel.

d) Type of grinding machine used. Production machines should generally be heavy and robust and will therefore take a softer grade of wheel than a lighter machine carrying out the same process.

Variable Factors

e) Grinding wheel speed.

f) Work speed.

g) Condition of the grinding machine, and particularly the state of the wheel spindle bearing.

h) Skill of the operator. If the machine is operator controlled, studies have shown that a man being paid on a piece-work basis will require a harder wheel than a man being paid on a day-work basis (i.e., paid by the hour). This is because in the first case the man will tend to force the cut harder in an effort to maintain a high production rate.

The number of factors which must be considered when choosing a grinding wheel for a particular operation, means that selecting the correct wheel becomes a confusing and difficult job. It is rather a specialist activity, and grinding wheel manufacturers provide a technical service of information and advice.

In order to see why the grinding process is an efficient metal removal process for roughing or finishing operations, we will examine the cutting action of a grinding wheel.

Cutting Action of a Grinding Wheel

The abrasive grains cut, not rub, small chips from the workpiece material. In effect the wheel presents a series of multi-point cutting tools to the work. The size of the chips removed depends upon the grit size used in the wheel. A fine grit should be used for finishing operations as it will remove finer chips and leave a smoother surface finish.

The ideal cutting action is when the work is wearing away the bond at the same speed as the grains are dulled. Hence the dulled grains are torn away exposing fresh grains. In this way the wheel is 'self sharpening'. The cutting life of an abrasive grit is comparatively short but because of this self sharpening action, repeated dressing should be unnecessary.

Figures 8.1 and 8.2 show the forces involved during a grinding operation.

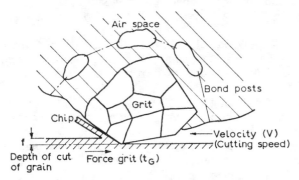

Fig 8.1   Cutting Action of an Individual Abrasive Grit.

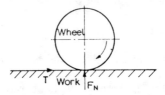

Fig 8.2   Surface Grinding Operation.

Referring to Fig 8.2, T and F_N are components of the forces acting on all the grits in contact with the workpiece.

It follows that the forces acting on the grit will increase as the cutting edge deteriorates, as is true for single point cutting tools. As the component forces T and F_N increase so the force/grit must increase.

When the force/grit becomes high enough to fracture the bond posts (Fig 8.1), then the dull grit will break away, and a new sharp grit will be exposed, giving the self sharpening action. It can be seen then that if too hard a grade is chosen for the wheel bond, the bond will not release the abrasive grains when they are dulled. This leads to local heating of the work surface, causing burn marks and a poor finish. The wheel will glaze.

It is not always practical when using expensive wheels on production machines to change the wheel for one of a more suitable grade in order to overcome the difficulties just described. However, the value of the force/grit t_G can be raised by increasing the work speed (using the same wheel), and this leads to the self sharpening action being maintained. Modern machines have a large range of work speeds, and on hydraulic machines the range is often infinitely variable. The effect of changing work speed should be fully exploited, as it is not always possible or practicable to change the wheel speed or depth of cut f, in order to change the cutting conditions.

As one would expect from what is primarily a finishing process, the size of chip removed by a single grain is minute. Consider the well known Guest Theory of grinding, as applied to an external cylindrical grinding operation.

Let V = Wheel velocity (m/s) N = No. of grains/unit length of wheel periphery. (Measured by rolling a wheel upon smoked glass and counting the marks left under a microscope.)

  v = Work velocity (m/s)

  d = Work diameter (mm)

  D = Wheel diameter (mm) L = Length of arc of contact between wheel and work.

  f = Radial in-feed (mm)

  t = Max^m chip thickness/grain (mm)

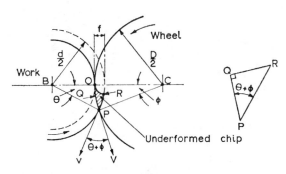

Fig 8.3 Geometry of Chip Formation in External Cylindrical Grinding

With reference to Fig 8.3, consider the action of a single grain at O and a point on the work at P. While the grain moves through the *arc of contact OP*, owing to the difference in the values of V and v, the work point will move to R and a chip OPR will be removed. QR = Maximum grain depth of cut. There are a number of grains operating along OP at any one time.

$$\text{Pitch of grains} = \frac{1}{N}$$

let T = Time taken for a grain (followed by NL grains), to move from O to P and a point on the work to move from P to R.

Then $OP = VT$, or $L = VT$. Therefore $T = \dfrac{L}{V}$

and $PR = vT$

Now $QR = PR \sin(\theta + \varphi)$. Therefore $QR = vT \sin(\theta + \varphi)$

$$\text{and} \quad t = \frac{QR}{NL} = \frac{vT}{NL} \sin(\theta + \varphi)$$

$$\text{Therefore} \quad t = \frac{v}{NV} \sin(\theta + \varphi) \quad\quad\dots\dots(1$$

Converting (1 into terms of D and f)
Consider $\triangle COBP$ (using Cosine Rule)

$$\left(\frac{D}{2} + \frac{d}{2} - f\right)^2 = \left(\frac{D}{2}\right)^2 + \left(\frac{d}{2}\right)^2 - 2 \frac{D}{2} \cdot \frac{d}{2} \cos(180 - \theta + \varphi)$$

f is small compared with D and d, hence expanding and omitting f^2

$$\therefore \frac{Dd}{2} - Df - df = \frac{Dd}{2} \cos(\theta + \varphi)$$

$$\therefore \cos(\theta + \varphi) = 1 - 2f\left(\frac{D + d}{Dd}\right)$$

Using $\sin^2(\theta + \varphi) = 1 - \cos^2(\theta + \varphi)$

$$\therefore \sin(\theta + \varphi) = 2 \cdot \sqrt{\frac{D + d}{Dd}} \cdot f$$

Hence substituting for $\sin(\theta + \varphi)$ in (1

$$\therefore t = \frac{2v}{NV} \sqrt{\frac{D + d}{Dd}} \cdot f \quad \text{for external cylindrical grinding.}$$

$$\text{and } t = \frac{2v}{NV} \sqrt{\frac{D - d}{Dd}} \cdot f \quad \text{for internal cylindrical grinding.}$$

If d is considered $= \infty$ for surface grinding then $t = \frac{2v}{NV} \sqrt{\frac{f}{D}}$

for surface grinding and $f =$ depth of cut (mm).

Example 8.1
Calculate the maximum chip thickness t for an external cylindrical finishing grinding operation if work diameter $= 38 \cdot 10$ mm, grinding wheel diameter $= 0 \cdot 305$ m, wheel velocity $= 25$ m/s, work velocity $= 0 \cdot 90$ m/s, the infeed $= 0 \cdot 13$ mm, and the abrasive grains are $0 \cdot 76$ mm apart.

Solution

$$\frac{1}{N} = 0 \cdot 76$$

$$t = \frac{2v}{NV} \sqrt{\frac{(D + d)}{Dd}} \cdot f = \frac{2 \times 900 \times 0 \cdot 76}{25\,000} \sqrt{\frac{(305 + 38 \cdot 1)}{(305 \times 38 \cdot 1)}} \times 0 \cdot 13$$

$$= 0 \cdot 054\,7 \sqrt{\left(\frac{343 \cdot 1}{11\,600}\right)} 0 \cdot 13 = 0 \cdot 054\,7 \times 0 \cdot 006\,2 = 0 \cdot 000\,31 \text{ mm.}$$

As stated earlier, with a finishing cut, the chip thickness is minute leading to a high degree of finish.

Although larger amounts of metal can be removed by grinding, it is not an economic proposition to do so on high production work. This can be demonstrated by examining the power required to remove metal by grinding.

Power required for grinding (P)

$$P = TV \text{ where } P = \text{power (W)}$$

T = tangential force upon the wheel (N) (see Fig 8.2)

V = wheel speed (m/s)

A good parameter by which to compare the economics of metal cutting processes is the energy output per mm³ of metal removed per second, which we will denote by the symbol K, where

$$K = \text{J/mm}^3 = \frac{P}{w}$$

Where w = volume of metal removed/second

$= f \times$ width of cut $\times$ work speed

Example 8.2

In a surface grinding operation, the tangential force = 90N, wheel speed = 25 m/s, work speed = 0·15 m/s, width of cut = 12·7 mm, and the depth of cut = 0·015 mm. Calculate the power at the spindle, and also the value of K.

Solution

$$P = TV = 90 \times 25 = 2250 \text{ W or } 2·25 \text{ kW}$$

$$K = \frac{P}{w} = \frac{2250}{0·015 \times 12·7 \times 150} = 83·3 \text{ J/mm}^3$$

This is a high energy output for the unit volume of metal removed per second and demonstrates how uneconomic it is to remove large amounts of metal by grinding.

8.2 MACHINE LAPPING

Hand lapping as a means of improving the surface finish of work such as gauges, has been practised for many years. Here, however, we shall consider the machine lapping of work required in larger quantities. Lapping is a finishing operation which results in three major refinements in the workpiece: (1) extreme accuracy of dimension, (2) correction of minor imperfections of geometric shape, and (3) refinement of surface finish. There is no distortion in the lapping process as it does not require the use of magnetic chucks or other holding or clamping devices, and less heat is generated than in other finishing operations.

In any type of lapping the results depend upon:

a) *Type of lap material.* Cast iron is the best, but also bronze, brass, lead and soft steel are used. In any event the lap should be softer than the workpiece so that the abrasive compound becomes embedded in the lap.

b) *Type of lapping medium or abrasive material.* Silicon carbide is used for rapid stock removal, and aluminium oxide for improved surface finish. The lapping compound (grit and oil) is best purchased from the manufacturers. A grit size from 100 to 1 000 is used depending upon the work being done. For example, 300 grit size could be used for primary lapping, followed by finish lapping or polishing using 800 grit size.

c) *Speed and pressure of the lapping motion.* Speeds between 1·50 m/s and 4·0 m/s are used whether the lap is rotating over the work or vice versa.

A pressure of 0·007 – 0·02 N/mm² (soft materials) up to 0·07 N/mm² (hard materials) is satisfactory. Excessive pressure causes rapid break-down and may score the workpiece.

d) *Material to be lapped.* This will affect the surface finish. Softer and non-ferrous materials require finer grit size to produce surface finishes comparable with those produced on hardened steel.

Lapping machines are used for improving *external cylindrical surfaces,* or *flat surfaces.* We will consider each in turn.

External, Cylindrical Machine Lapping

This is usually a vertical spindle machine carrying one upper stationary lap and one rotating lower lap. The upper lap is free to float and rest upon the work which rides upon the face of the lower lap. Pressure is applied by gravity. The work is held loosely in a work guide or holder; hence it follows a random, complicated path between the lap faces. In cylindrical lapping the work propels the work holder. Both lap faces must be truly flat. A diagram of the process is shown at Fig 8.4.

The final size of the work is held to close limits because the rate of stock removal decreases as more and more workpieces clean up. 0·001 mm accuracy of dimension can be held on production, with a surface quality value of 0·025 μm R_a.

Flat Machine Lapping

Similar type of machine to that used for external cylindrical lapping with some differences. The work holders propel the work in this case instead of vice versa. A diagram of a flat lapping machine is shown at Fig 8.5.

The rotating driving spindle may give a friction drive to the work holders, or a positive drive may be given through gear teeth on the periphery of the spindle and work holders. The configuration is then rather like planetary gearing. The driving spindle rotates at a different speed than the lower lap, and the motion given to the work holder causes the work to cover the entire lap surface in a random pattern. The resultant flatness, parallelism of faces, and uniformity of dimension is superior to that which can be obtained by any other finishing process, including hand lapping. Optical flatness can be obtained giving a *wringing* surface as found on slip gauges.

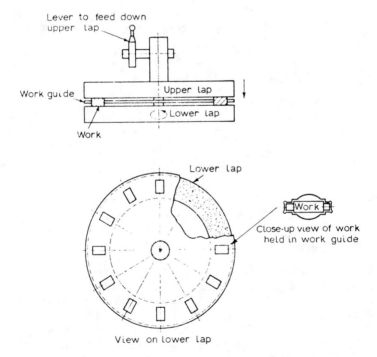

Fig 8.4 External Cylindrical Machine Lapping.

On some of the simpler machines of this type, no top lap is used. One component face at a time is lapped, pressure between the work and lap being given by the weight of the work; sometimes supplemented by weights resting on the work.

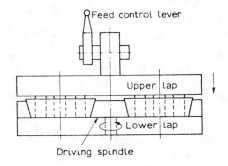

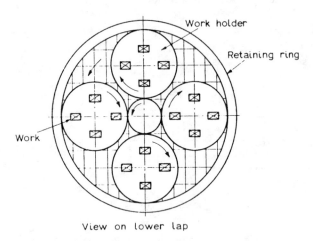

Fig 8.5 Flat Lapping Machine.

8.3 HONING

Internal cylindrical surfaces of high quality are not usually finished by internal lapping, but by honing. Honing is a wet cutting process which removes metal in the workpiece by means of revolving, reciprocating abrasive sticks (up to eight in number) carried in a suitable holder. It is usually applied to internal cylindrical surfaces but special machines have been developed for honing external surfaces. Honing achieves four things:

1) Removes metal:
   Up to 0·5 mm for primary honing.
   Up to 0·01 mm for secondary or mirror honing.
2) Generates size:
   Limit of accuracy on holes up to 50 mm dia. is 0·005 mm.
   Limit of accuracy on holes between 50 mm and 150 mm dia. is 0·01 mm.

Vertical machines have been designed for work up to 500 mm dia. Horizontal machines have been designed for work up to 8 m in length.

3) Generates roundness and straightness.

Accurate roundness is generated by the rotary motion of the hone, and the controlled, positive expansion of abrasive sticks. Accurate straightness is generated by the reciprocating motion of the hone. The sticks should be long enough to correct any waviness in the bore, and should be set to emerge about one-third of their length either side of the hole where possible.

4) Produces any required surface finish. Depends upon the combination of speeds and radial feeds of the sticks and grit sizes of sticks.

Primary honing: 80 – 180 grit size.

Secondary honing: 300 – 500 grit size.

Surface finishes of 0·05 μm R_a can be achieved.

It is important to note that the honing tool will follow the axis of the original hole, so the tool or the fixture must be free to float. A sketch of a honing spindle is shown at Fig 8.6.

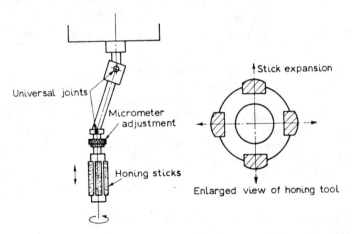

Fig 8.6 Honing Spindle.

The abrasive sticks are identical in structure to grinding wheels When the sticks are removing metal the hole is flooded with a lubricant such as paraffin, throughout the operation. The hone rotates at 0·5 m/s to 2·5 m/s and reciprocates at 0·2 m/s to 0·5 m/s. The sticks are expanded while honing takes place if required, by micrometer controlled, mechanical or hydraulic means. As shown in Fig 8.6, the universal joints make it unnecessary to line up the hole and hone axes precisely, as the hone is self centring.

The honing process uses a greater number of cutting points than are found in most abrading processes, e.g., honing a 76 mm dia. × 200 mm long bore, six 150 grit abrasive sticks would have 140 × the abrasive contact area of an internal grinding wheel, and 98 000 shearing contacts against 46 contacting grains of a grinding wheel. This distribution of cutting points over a large contact area also reduces heat generated during cutting. Consider the cutting action of one honing stick, as shown in Fig 8.7.

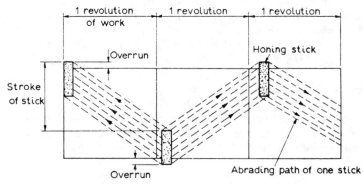

Fig 8.7   Schematic sketch of the path of movement of one Honing Stick.

The sketch shows a ratio of revolving movements to reciprocating movements as 2 +/1. In practice this is kept to an odd-ratio, such as 3 +/1. With the selection possible in numbers of sticks and ratios of movements, the paths of the sticks will cross each other many times during the operation. A wide variety in the angle of the cross hatch of the spiral paths will be obtained, and the whole internal surface will be abraded many times by the abrasive grits in each stick.

On hydraulic production machines, several spindles may be fitted to one machine and automatic size control can be incorporated if required. Its functions are:
1) Stop radial feed-out of tool when bore is size.
2) Collapse tool.
3) Withdraw tool from workpiece.
The first function may be carried out using electrical contacts, which are set to a master gauge, to close when the job is size. This energises a solenoid which in turn operates a hydraulic valve controlling the stick feed-out mechanism.

8.4 SUPERFINISHING

This is a machine finishing process, similar in principle to external honing, in that the work surface is generated to its true geometric shape by abrading. Fig 8.8 shows the principle of operation of the process.

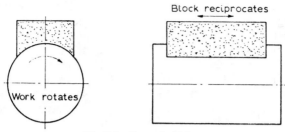

Fig 8.8 Superfinishing.

Like honing, the abrasive block (of appropriate form), reciprocates across the face of the work with a similar amount of over-run at each end of the stroke. At the same time the work rotates about its axis. These two motions give a high degree of geometric accuracy to the workpiece.

The contact area between the abrasive face and the work is greater than in honing, giving more cutting points, and even less pressure/grit and hence less heat/grit. The work is flooded with lubricant during cutting. The resultant surface finish cannot be surpassed by any other process and can be produced to the order of a R_a value of 0·0125 μm.

8.5 DIAMOND MACHINING

Diamond is the hardest cutting tool material used in engineering and has several specialized applications. For example, diamonds are used in the abrading processes, as a means of truing or forming the faces of grinding wheels, as the abrasive grit in abrasive wheels used for dressing sintered carbide tools, or as a loose abrasive grit used as a lapping medium for polishing carbide dies. Diamonds themselves are sometimes used as a die material for wiredrawing. In this section, however, we wish to consider the diamond used as a single point cutting tool for finishing operations. Diamond machining is usually confined to finish boring or finish turning.

Natural diamonds are used for industrial purposes, although synthetic diamond can now be obtained mainly in the form of grit. In both forms it is still a very expensive material. It has several outstanding properties which make it suitable for a cutting tool material, these being great hardness and strength, low friction, high thermal conductivity and a low coefficient of expansion. Diamond is at the top of the M o h Hardness scale having a value of 10, and of course its resistance to wear is outstanding. With these properties, a diamond cutting tool is capable of machining the hardest of metals yet known. However, the diamonds are so small, and in addition extremely brittle, that they are mostly used upon the softer materials. They are most suitable and productive when used for taking fine finishing cuts, at very high cutting speeds, on the non-ferrous metals. They have also been used successfully for machining the very abrasive polymers, ceramics, amd also rubber.

Due to the extreme brittleness of diamonds precautions must be taken to ensure that rigidity is at a maximum and vibration is at a minimum when they are being used as cutting tools. These principles are identical to those which must be observed when using tungsten carbide cutting tools, if the finished results are to be satisfactory. The tool must be soundly clamped with the minimum of overhang. Special machines have been designed for diamond boring and turning, and they have the following general features:

i) Large, plain bearings supporting the main spindle giving a minimum of vibration.

ii) Robust machine bed and structure giving maximum rigidity.

iii) The drive to the main spindle being taken through endless belts, instead of through gearing. Again, this is in order to reduce vibration. Hydraulic precision boring machines are used with great success for diamond boring, being vibration free in operation.

Diamonds, being so small, are used in the form of tips which are either clamped or cold set into a steel shank. The shank can then be used in a tool holder, such as a micro-boring head (micrometer controlled adjustable boring head), for example. The tips can be obtained commercially with the cutting edge either radiused, or in the form of a series of flats. See Fig 8.9.

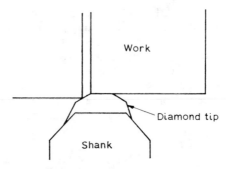

Fig 8.9 Diamond Tip.

With this tip shape, the cutting edges are to the order of 0·4 mm long. One edge should be set accurately parallel to the work axis as shown, a setting microscope being useful for this purpose. When one edge is eventually dulled after use, the tip is rotated to bring the next cutting edge into position. A diamond tool cannot be re-sharpened in the workshop, but must be returned to the supplier for polishing. It is also essential that the tool is set accurately on the work centre line.

The following information will give a guide to the speeds and feeds used for diamond machining.

Feed. Feeds must be light, to the order of 0·01 mm to 0·04 mm/revolution, anything higher being considered a heavy cut.

Speed. Cutting speeds from 5 m/s to 50 m/s have been used, and should be kept as high as the machine and work material will allow.

Depth of cut. This will always be as little as possible as it is a finishing operation. 0·15 mm can be taken to be a heavy cut.

Cutting forces will be light.

Two highly successful diamond machining operations that are carried out at high production rates are:

1) Turning of the outside surface of aluminium alloy pistons.
2) Boring of white metal bearing liners.

A highly polished surface finish can be obtained which can be said to be of super-finish standard, and is comparable with the results of any other finishing process.

Exercises 8

1. Describe the cutting action of a grinding wheel, referring to the Guest theory of grinding where appropriate.

2. The following code is marked upon a grinding wheel: A 36 M 5 B. State the meaning of each symbol, and outline in detail the significance of each factor, denoted by a symbol, on the cutting efficiency of the grinding wheel.

3. With the aid of sketches, show the general configuration and main features of the following types of grinding machines: (a) surface, (b) external cylindrical, and (c) internal cylindrical.

Why do you think that a universal grinding machine would not be a suitable production grinding machine?

4. Compare the centreless type cylindrical grinding process with the more orthodox centre type grinding process as a means of producing ground cylindrical surfaces on work required in large quantities.

5. Three workpieces made of the same type of material are being finished by (a) surface grinding, (b) external cylindrical grinding, and (c) internal cylindrical grinding respectively. In each case, all the grinding parameters such as infeed, work speed, type of wheel, etc., are identical, with the exception of grinding wheel speed. In each case, this is substantially different. Why?

6. In a surface grinding operation carried out on a horizontal spindle machine, the following information is known:

Grinding wheel diameter = 300 mm
Grinding wheel velocity = 25 m/s
Table speed = 0·2 m/s
Depth of cut = 0·02 mm
Width of cut = 10 mm
Grain spacing = 1 mm
Tangential force = 90 N

Calculate (a) the power at the spindle in kW, (b) the metal removal rate in J/mm³, (c) maximum chip thickness in mm.

(Ans. (a) 2·25 kW, (b) 56·3 J/mm³, (c) 0·000 37 mm.)

7. Describe the principle of lapping. Using diagrams show how it is applied to the machine lapping of external cylindrical surfaces, and flat surfaces.

8. By analyzing the movement of the paths of the abrasive sticks, show how honing improves the geometry and finish of a cylindrical hole.

9. What is diamond machining? Show how it could be applied to the micro-boring of an engine conn-rod big end bearing liner, in large quantities using a duplex, horizontal spindle machine. Sketch a pneumatic fixture for holding the conn-rods.

10. Compare grinding, lapping, honing, superfinishing and diamond machining as finishing operations, giving the expected range of surface finish values for each process.

Further Reading

1) Colvin and Stanley. *Grinding Practice*. McGraw-Hill Publishing Co. Ltd.

2) Gisbrook H. 'Precision Grinding Research.' *Journal of the Institute of Production Engineers*, May 1960.

3) Wills H. J. and Ingram H. J. 'Fine Grinding and Lapping.' *Machinery's* Yellow Back Series, No. 19.

4) Allan G. A. and Sutherland K. H. 'A Preliminary Study of the Lapping Process.' *Journal of the Institute of Production Engineers*.

5) Brunton J. H. 'Diamonds as an Engineering Material.' *Journal of Institute of Mechanical Engineers*, January 1966.

6) Scriven A.' 'Synthetic diamonds for industrial tools.' *Journal of the Institute of Mechanical Engineers*.

7) König W. 'New methods of cutting and grinding.' *Journal of the Institute of Production Engineers*. December 1966.

CHAPTER 9

Automatic Control of Size

9.1 AUTO SIZING

AS MORE engineering processes are carried out by semi-automatic and fully automatic means, so the trend will be towards gauging or measuring the work produced by the same means. We must make a distinction here between gauging and measurement. Measurement reveals the size, and hence the error if any, of the workpiece; gauging can only check the work to see if it is correct, but does not reveal the extent of any error which may be present.

Gauging (or limit gauging, to give it its full name) is preferred for quantity production work because:

a) It is quicker than direct measurement, and often the magnitude of any workpiece error is not required; only the knowledge that it is acceptable or a reject.

b) It is often easier to design an automatic gauging system than an automatic measuring system.

Gauging has disadvantages which are:

a) It may be necessary when using statistical quality control techniques (see Chapter 14) to know the magnitude of component size variations. In such a case, gauging cannot be used and must be replaced by direct measurement.

b) Gauges themselves are subject to manufacturing error, or to errors created by wear in use. This latter point is of particular importance in some automatic gauging systems.

Bearing in mind the distinction between measurement and gauging, automatic control of workpiece size during machining is known as *in-process gauging, in-process measurement, autosizing* or *automatic inspection*. These terms are used quite loosely to describe the process under discussion in this chapter, and one should be aware of the imprecise terminology. We shall be concerned only with devices which measure or gauge the size and geometric features of a workpiece.

Auto sizing, which is short for automatic control of size, is a subject with a great variety of methods of application. We will try to introduce some order into the subject by categorizing the different types of auto sizing systems. Firstly, the system may be *open loop* or *closed loop*, these terms being explained in Section 6.1.

Open Loop
In this case the work will be automatically gauged or measured during the machining of the component (hence the term, in-process). The size will be presented to the operator in such a way that he can then set the machine to produce a finishing cut, confident that the final workpiece size will be within the limits specified.

This is in effect a semi-automatic system. It has been applied mainly to the control of diameter, particularly on grinding machines. Commercial systems are available which enable ground outside diameters to be controlled to within $\pm\, 0\cdot001$ mm, which is generally more than adequate for quantity produced work, and special systems are available of even greater accuracy.

Closed Loop
In this case, the work is automatically measured during machining as before. The system incorporates *feed back* such that the variations in the work size above the specified limit are corrected by the automatic resetting of the machine. The final work size is achieved by automatic monitoring of variation in workpiece size as the removal of surplus material by machining takes place. This is a fully automatic, inspection control system.

Secondly, the system may be *direct* or *indirect*.

Direct
In this case the workpiece is gauged or measured directly whilst being machined, i.e., the gauging or measuring head is applied directly to the work surface which is being machined. The most commonly used systems applied to external cylindrical surfaces employ calipers as a means of contacting the work surface.

Indirect
In this case, the control of workpiece size is attained indirectly either by monitoring the position of the cutting tool slide, or by monitoring the position of the tip of the cutting tool. Ideally these systems should conform to the Principle of Alignment (Abbé Principle) such that the scale axis lies along the same axis as the tool in-feed is acting along. In practice this is usually impossible.

The diffraction grating, monitoring system for numerical controlled machine tools, described under Section 6.7 is a closed loop, indirect auto sizing system in which the finished size of the workpiece is controlled to close limits of accuracy. Theoretically, direct measurement should give the most accurate results. Whichever system is used, the accuracy obtained from the auto sizing unit depends upon the inherent errors in the machine tool, the type of machining process being carried out, and the normal errors bound up in the act of gauging or measuring with all its associated parameters, such as wear, temperature differences, etc. Each section in this chapter will be concerned with a common auto sizing device which will be either open loop or closed loop, and either direct or indirect in principle. We leave it to the reader to decide into which category each example falls.

9.2 MECHANICAL CALIPER FOR A TURNING OPERATION

Auto sizing of turned parts upon production is not common. Fine tolerances are not usually required on quantity production turning operations, so that occasional gauging is all that is required. Also the semi and fully automatic lathes described in Chapter 4 make use of cams or stops to control work size, and the additional use of in-process gauging systems cannot be justified. All this means that auto sizing is most commonly applied to finishing operations, and grinding in particular. However, there is a small proportion of production turning work carried out where high accuracy is required, and scrap rates are proportionally higher than average. In such class of work, in-process measurement could be economically justified.

When applying such a system to turning, several practical difficulties exist. For example, stock removal is high, hence the system must have a large enough range to cope with the comparatively wide variation between the largest and smallest diameter. Surface finish is rougher than that produced by finishing processes, swarf control is a problem, and the inherent accuracy of the machine tool is less. The instrument used then, must be robust and simple, with a wide range and capable of detecting size variation in work being turned to an accuracy of $\pm\,0.01$ mm, say.

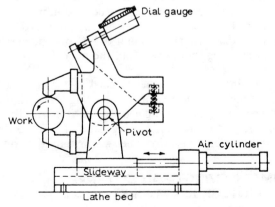

Fig 9.1 In-process Measurement of Turned Work.

An auto sizing device for turning operations is shown at Fig 9.1.

This instrument makes use of scissor calipers having a 1 – 1 ratio, and solid jaws. Calipers are used frequently for direct application, being mechanically simple and trouble free. In order for a scissors caliper to have maximum accuracy, the geometry of a similar triangle must be maintained for the instrument throughout its whole range of measurement. When used upon a centre lathe or turret lathe, it must be mounted on the back of the machine being adjustable for position along the length of the bed. Like much in-process measuring equipment, it can interfere with tool access to the work. A simple air cylinder, or some other means can be used to move the instrument forward to straddle the work during machining, and to retract it during workpiece unloading.

Calipers can be used with any type of displacement transducer, and Fig 9.1 shows the movement of the caliper jaws being recorded on a dial gauge. A dial gauge can be classed as a type of mechanical transducer having good resolution and a wide range. The accuracy and repeatability will depend upon the quality of gauge used, but the lathe operator should be able to read the variation in work diameter shown upon the gauge quite easily. The device is in fact a simple comparator, and therefore must be set to read zero on a rotating master gauge of the required mean diameter before being used. The operator will control the amount of in-feed applied to the tool during cutting, his control of the situation being determined by the information he receives from the caliper gauge. This illustrates the big advantage of direct in-process measurement, which is that the size is controlled during manufacture, instead of being checked after manufacture, and is independent of tool wear. Therefore, scrap work should be minimal as the work sizes will not start to drift outside the upper limit due to tool wear, as is the case when no auto sizing equipment is used.

Finally, one other fundamental point of principle must be mentioned. A caliper gauge of the type described can be used in two ways. Firstly, it can be fastened in a fixed position on the lathe bed such that it records the work diameter at one point only. This one diameter must then be taken as being indicative of all the diameters along the work length.

Secondly, the instrument can be fastened to the lathe saddle such that it traverses along the work length, recording the work diameter at every point along its length. In order to do this it must follow the tool, and will operate just behind the cutting zone. In this position it will receive maximum interference from the swarf removed during cutting.

9.3 PNEUMATIC SIZING OF AN EXTERNAL CYLINDRICAL GROUND WORK

The grinding of outside diameters lends itself more easily to autosizing than does the turning of outside diameters. The reason is that grinding involves much less metal removal with little swarf or cuttings, the surface finish is finer, and being a finishing process it lends itself better to the use of sophisticated measuring devices. Before looking at a pneumatically controlled automatic sizing device, let us examine the principle of operation of low pressure air gauging.

Air Gauging

Air gauging, using low pressure air (or more recently high pressure air), is a very popular medium for automatic inspection of ground components and can be applied in a variety of ways. It is ideal for providing feed back signals in a closed loop system in order to regulate the machine feed control according to the varying work size. This is shown at Fig 9.2.

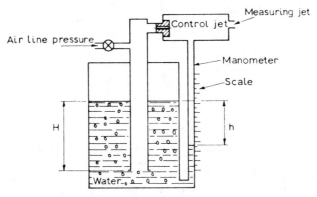

Fig 9.2 Principle of Air Gauging.

Figure 9.2 shows a metal vessel containing water in which is housed an open ended dip tube reaching nearly to the bottom of the vessel. Air is fed to the upper part of the dip tube at a pressure p of about 0.14 N/mm^2. The arrangement shown allows air to pass out through the control jet at a constant pressure (which is an essential requirement of accurate air gauging) although the incoming air line pressure may fluctuate. Air in the chamber behind the control jet is kept constant and equal to the head of water H forced down the dip tube, excess air bubbling out from the bottom of the tube and escaping freely to atmosphere. So long as air line pressure p is higher than H, air will escape by bubbling from the tube, and while this happens the control jet air pressure will be constant. Fluctuations in air line pressure will merely cause a lesser or greater amount of bubbling.

Air at a controlled pressure passes through the control jet at pressure H, and out through the orifice of the measuring jet. A manometer is placed with its inlet between the two jets. Variations in the flow of air through the measuring jet (due to the jet orifice being open or partially closed) cause a variation in the pressure *h*, which varies the height of the water in the manometer tube. Complete closure of the jet depresses the manometer water level to the bottom. When used as a measuring instrument, variations in the flow of air through the measuring jet are caused by variations in size of work being measured. If a large calibrated scale is set beside the manometer tube, then variations in work size can be read directly from this scale.

Air Controlled Automatic Sizing
One example of air gauging techniques applied to auto sizing is shown at Fig 9.3 which illustrates a radius reading device applied to external cylindrical grinding.

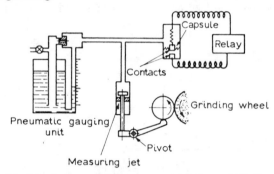

Fig 9.3   In-process Measurement of Ground Work.

Figure 9.3 shows a pivoting arm making contact with the work. As the work is reduced in size, so the arm will pivot and cause the attached valve to slowly close the orifice of the measuring jet. This in turn increases the air pressure '*h*' (see Fig 9.2) in the manometer, and a direct reading of size variation can be taken from the manometer scale by the operator.

In addition, the system can be increased to give fully automatic control as shown in the diagram. Increasing back pressure of the air from the measuring jet will also cause the pressure sensitive capsule to operate in the cut-out unit, thus closing the two electrical pre-set contacts. When they close, the circuit is closed and hence the relay operates. The relay controls the action of the grinding wheel in-feed mechanism, and when the relay operates the in-feed stops, and the component is size.

Radial reading devices of the type just described are very useful for automatic sizing, but their accuracy is affected by any change in work axis, unlike diameter reading devices such as calipers. Thermal effects can cause a change of position of work axis, and this is why high accuracy machine tools such as grinding machines are sometimes run for a period before cutting commences, in order to warm up the spindle and bearing assembly.

Measuring devices which use arms which carry tips making contact with the work, will be subject to tip wear after long use. In order to reduce wear to a minimum the tips are often faced with stellite, carbide, or even diamond.

9.4 IN-PROCESS GAUGING OF INTERNAL CYLINDRICAL GROUND WORK

In-process gauging has been successfully applied in various ways to internal ground work, both centreless and orthodox. Again, air gauging lends itself well to this process using internal calipers, as the work contacting medium, or a single, radius reading contact. (These devices are suitable for bores above 40 mm diameter.) Machines used for this type of operation usually incorporate automatic wheel dressing at the end of the rough grinding operation. Therefore, it is usual to incorporate two stage controllers for internal grinding machines which have the following cycle:

a) Automatically inspect bore during rough grinding.

b) Activate wheel dressing when bore is a size equal to finished diameter minus final grinding allowance.

c) Automatically inspect bore during finish grinding.

d) Stop grinding operation when bore is finished size.

The two devices described in Sections 9·2 and 9·3 are measuring devices. In order to show a variety of principles of operation, we will describe a gauging device in this section. Figure 9.4 shows the in-process gauging of internally ground work.

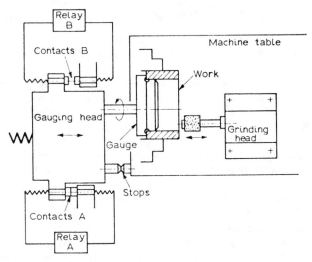

Fig 9.4 In-process Gauging of Ground Work.

The equipment shown in diagrammatic form at Fig 9.4 could be electro-hydraulically operated and is a two stage controller. As the machine table carrying the wheel head starts to traverse, a stop on the end of the table presses against the gauging head stop, moving the gauging head and hence the gauge, away from the work as the grinding wheel traverses the length of the work. This allows the wheel to over-run the back of the hole without striking the gauge. As the wheel moves to the right, out of the hole, the gauge is presented to the back of the hole. The plug gauge has two diameters equal to the roughing size and finished size of the hole respectively.

When the front (roughing diameter) part of the gauge eventually enters the hole, contacts A close, and cause relay A to operate. This is the stage of the cycle shown at Fig 9.4. The table now short strokes automatically, the diamond drops into position, and the wheel is dressed. The wheel starts regrinding at a diminished feed, again the gauge trying to enter the hole at each stroke. The gauge does eventually enter when the hole is size, and thus contacts B will close. Relay B is activated, the table returns to the rest position and grinding stops. The gauge, which rotates with the work during the operation, stops and retracts from the work under hydraulic control. The work could then be automatically unloaded from the chuck, and the chuck re-loaded.

The reader will appreciate, after consideration of Section 8.1, that an internal grinding wheel must be softer than an external grinding wheel used for an equivalent operation. Therefore, the internal grinding wheel requires frequent re-dressing due to the greater wear rate.

9.5 PNEUMATIC SLIDE POSITION MEASURING DEVICE

Before examining a typical slide monitoring unit which can be used for indirect auto sizing, let us look at the subject of slide positioning in general.

Firstly, the accuracy of operation of any general purpose machine tool can be increased by improving the method of indicating slide position. Backlash is a problem with leadscrew control, and often micrometer dials on leadscrews are poorly calibrated. It may be noted that an improvement in accuracy is brought about by the use of lathe micro-dials which are calibrated to record the diametral effect on the workpiece rather than radial in-feed. Commercial units are available which can be fitted to existing machines, and of course on toolroom machine tools such as jig borers, highly accurate slide monitoring units are built into the machine.

Secondly, if the operator on a production machine is to adjust the tool slide position as a result of information received from a direct, open loop, auto sizing device, then he will need a more accurate slide positioner than is usually provided. Indeed, it will be necessary to have the means of positioning the slide with twice the accuracy of the equipment measuring the work diameter.

We can now look at the principle of operation of an indirect auto sizing device, which is usually termed a feed gauge. It is initially set to a standard gauge of the required size, and the operator must then set the slide to this datum position for the machining of each component. Therefore, no account is taken of tool wear, and although the slide is set accurately to the datum position each time, the component size could be gradually increasing without the auto sizer revealing it. This is a disadvantage of these units, and post process inspection will be necessary.

Figure 9.5 shows a pneumatic, slide positioning device suitable for use on a production grinding machine, or lathe. Remember that any error due to this feed gauge will produce twice that error upon a round workpiece.

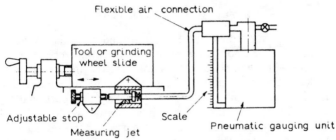

Fig 9.5   Pneumatically operated Slide Positioning Unit.

The feed gauge illustrated at Fig 9.5 is simply an air comparator adapted to a special purpose, like the air caliper shown in Section 9.3. In this case the leadscrew is used only as a means of moving the slide along its slideway, and the readings on the leadscrew dial are not used. An adjustable stop is fixed to the machine frame and is preset to suit the master gauge. The measuring jet head is fixed to the slide whose position is to be recorded. As the slide is moved, and reduction of the work diameter takes place, the valve pressed against the adjustable stop will gradually uncover the measuring jet orifice. So the water level in the manometer tube will gradually change until the datum point on the scale is reached.

The range of movement of the device described is very wide compared with a pneumatic caliper device which has a limited range of movement.

9.6 DIGITAL SLIDE POSITION MEASURING DEVICE

High class, electrical monitoring systems are available which can easily be fitted to any production machine slide. The measuring unit is a linear displacement transducer (referred to in Section 6.7) which is usually some type of potentiometer, such as an inductive or a capacitive type, these two being very similar in principle. The measurement of slide movement given by the measuring unit is displayed in digital form (see Fig 2.3) on an indicating meter.

Figure 9.6 shows in very simple terms the principle of a capacitive potentiometer.

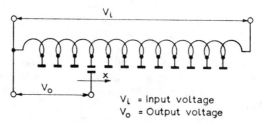

V_i = Input voltage
V_o = Output voltage

Fig 9.6   Capacitive Potentiometer.

A slider can be moved along a former around which is wound many turns of wire. Tapping points are taken from the wound coil as shown in Fig 9.6, with which the slider makes contact. The input voltage Vi is kept constant, and the output voltage Vo changes in steps when the slider is moved through a distance x, i.e., Vo is proportional to the distance moved by the slider. If the slider is attached to the machine slide then any movement of the slide is sensed by the transducer just described, and is recorded upon a digital unit of the type shown in Fig 9.7.

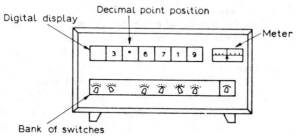

Fig 9.7   Digital Indicating Unit.

These units are made to different patterns by the manufacturers of high class equipment, such as E.M.I., S.I.P., and Ferranti, for example. They are designed to sound ergonomic principles as outlined in Section 2.2. Figure 9.7 shows a typical cabinet.

The numerical value shown by the display can be altered to any value by adjusting the individual switches under the display. An accurate zero setting can be obtained on the meter by use of the switch underneath the meter panel. Various techniques can be used with this type of equipment; for example the digital unit can record cross slide movement as radial in-feed, or diametral change in workpiece size. One useful method of operation is as follows:

a) Machine a diameter on a test piece of any convenient size leaving the tool in its set cutting position and accurately measure the diameter produced.

b) Display the numerical value of this diameter on the unit. (Note. The position of the decimal point is fixed.)

c) Set the meter to a zero reading. The tool position is now identified by the display.

d) Set the required value of the workpiece diameter to be manufactured on the display.

e) The tool can now be used to progressively machine each workpiece using one or more cuts of varying depth, and the workpiece is at the correct diameter when the meter reads zero.

Alternatively, if no meter is available, the tool can be set to the datum position, and the value set up on the display. The operator will then reach size on the workpiece when the required value is displayed on the unit. In this case he will have to remember the actual numerical value of the required diameter, rather than observe a zero setting position. Like the pneumatic feed gauge shown at Fig 9.5, this system is comparative and therefore requires a standard of known size to establish a datum position for the slide, and hence the tool.

These digital systems are less prone to operator error than any others, have a wide range (up to 500 mm in some cases), good resolution and repeatability, and very high accuracy (0·002 5 mm easily, and in more sophisticated versions may be 0·001 mm).

9.7 AUTO SIZING DEVICE FOR CENTRELESS GRINDING OPERATION

Let us consider a high output, precision process using a fully automatic machine, where high quality work is produced requiring 100% automatic inspection. The centreless grinding (followed by lapping) of gudgeon pins requiring high accuracy and finish would be an example. This represents the ultimate attainment in an automatic sizing and inspection device, because the in-feed of the wheel head is automatically controlled, the feed of work to the machine is automatically controlled and any rejects produced are segregated automatically. Such a system could be pneumatic-electric controlled, making use of the principles outlined in Sections 9.3 and 9.5. A schematic outline of an automatic control cycle suitable for a through feed, centreless grinding operation is shown at Fig 9.8.

Referring to Fig 9.8 we see that the auto sizing and inspection unit consists of five elements:

1) Feed hopper which controls the feed of the components to the grinding machine, and receives instructions in the form of electrical impulses from the controller.

2) Pneumatic gauging unit which contains pre-set inspection air jets which sense the work diameter, and send out impulses to the controller if the work diameter increases beyond the top limit, due to grinding wheel wear.

3) Electric controller which receives impulses from the gauging unit, and sends out appropriate instructions to the grinding wheel head feed mechanism to feed in.

4) Feed gauge (or wheel head slide positioning unit) which feeds back impulses to the controller so that the wheel slide is set to the correct position.

5) Reject chute which normally allows work through to the finished work hopper when the system is stable. When corrective measures are being taken to readjust the wheel slide position due to impulses received by the controller, instructions are sent to the chute which operates a deflector to divert work to a reject box. The small number of possible rejects involved can be occasionally collected and gauged by an inspector.

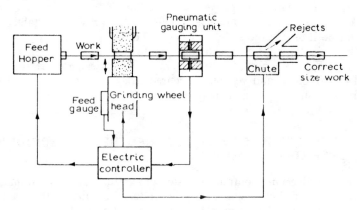

Fig 9.8   Automatic Control and Inspection Device for Centreless Grinding.

The feed hopper need only be filled with components from time to time, and the system will operate automatically. The control of the feed hopper is introduced to reduce the number of rejects even further, because the throughput of the machine is high and rejects could quickly build up in number during the wheel slide position correcting operation. Therefore, as soon as the controller receives a signal from the gauging unit (activating the auto sizing cycle), it sends out an instruction to the feed hopper to stop feeding work to the grinding machine. When the system is stable again the feed hopper will re-operate.

9.8 FRICTION ROLLERS

A less used, but interesting method of in-process measurement is based upon the use of rollers of known diameter, bearing upon the workpiece surface and being driven by friction. Hence the workpiece diameter can be determined by comparing the rate of rotation of the work with the rate of rotation of the roller, i.e., the workpiece diameter is obtained by a comparison of circumferences.

In practice, the rate of rotation of the rollers is obtained using an angular transducer. This may be a radial grating transmitting electric pulses which are counted by some sort of electrical counting device. The principle of operation here is identical to that of the linear diffraction gratings and pulse counters described in Section 6.7. In this case, of course, we are concerned with angular movement instead of linear movement. A count is made of the number of roller pulses transmitted during a certain number of workpiece revolutions, say 20 for example.

The workpiece diameter is then easily deduced by consideration of ratios of diameters and angular velocities and is given by: $D = \dfrac{dx}{NS}$

Where D = Diameter of workpiece.
$\quad\;\; d$ = Diameter of the roller.
$\quad\;\; x$ = Number of pulses counted during N revolutions of the workpiece.
$\quad\;\; S$ = Number of pulses produced/revolution of the roller. This is standard for the equipment used, and will depend upon the number of grating lines used.

The disadvantage of the system is that the monitoring of the change in work diameter is intermittent, and not continuous. The change in work size cannot be recorded for each workpiece revolution, but is recorded intermittently depending upon the value of N. If N is reduced to unity, in order to give continuous size assessment, then the resolution of the system is very low. The method is best applied to large diameter work rigid enough to withstand the high loads required between roller and work in order to ensure that no slip takes place. Roller slip, of course, would seriously affect the accuracy of the system.

An advantage of the use of friction rollers is that the range of the equipment is high compared to other direct, in-process measuring systems.

9.9 OPTICAL MEASUREMENT

Optical projection methods have been applied for some years to machine tools, and one thinks particularly of their use on precision form grinding machines. These are designed to facilitate the grinding of complex tool or die forms, and the position of the grinding wheel relative to the work can be seen through a microscope. Alternatively, it may be projected on to a screen, but there is a limit to the size which can be projected on to a screen of sensible size, using a reasonable magnification factor, say from $25\times$ to $100\times$.

Optical devices give a means of accurately positioning a tool tip, rather than positioning the slide carrying the tool as described in previous sections. In its simplest form, all that is required is a microscope of low magnification. The microscope can be mounted over the tool, which might be quite inaccessible to mechanical devices, and requires a target graticule eyepiece, and the means of fine adjustment. The set-up is shown at Fig 9.9.

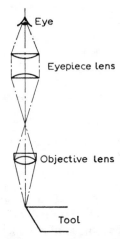

Fig 9.9   Optical Positioning of a Tool Tip.

If the objective lens magnifies $5\times$, and the eyepiece $10\times$, then the virtual image seen by the eye will appear to be $50\times$ full size. Again, this is a comparative unit, and the microscope must first be set to a master, and accurately positioned and locked in its datum position. The position of the tool tip is checked by the operator during production, and as the tool wears the position of the tool slide can be altered to reset the tool to the microscope graticule line, hence correcting the effect of tool wear and preventing the upward drift of the size of components. More sophisticated systems have been used which do not require a tool tip image to be positioned to a graticule, and give greatly increased accuracy.

Optical systems of direct, in-process measurement are not often used, and if they incorporate screen projection there is a limitation on workpiece size as mentioned earlier. The system shown in Fig 9.10 overcomes the difficulty of large diameter work, by using a split image device.

Figure 9.10 shows a microscope being used, but the work image could be projected on to a screen if required. This is more expensive but less tiring to the operator than looking through a microscope eyepiece. The internal reflecting prisms shown at Fig 9.10(a) provide the microscope with two virtual images in effect, which are the profiles of the opposite sides of the workpiece. These two images are seen combined (Fig 9.10(b)), and the width of the total image is proportional to the work diameter.

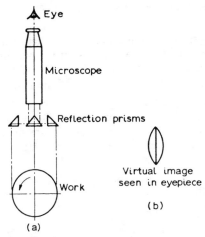

Fig 9.10 Optical In-process Measurement.

If the microscope graticule incorporates the means of indicating the work tolerance, then the tool slide position can be adjusted to suit as the operator observes the image approaching the limit. The system described requires the assistance of an operator in determining the work diameter as it is being processed, but more complex systems have been used which give an automatic reading of work diameter.

All optical systems have the disadvantage that they tend to be bulky, and therefore must be used well away from the work so as not to cause obstruction in the machining zone. They must also be kept clean and free from oil.

Exercises 9

1. Differentiate between in-process gauging and in-process measurement.

2. Show the difference between (a) direct and indirect, and (b) open loop and closed loop systems of in-process control of size.

3. Place each of the auto sizing systems described in this chapter into either category (a) or (b), Exercise 2 above.

4. By means of a diagram, show the principle of operation of an automatic, radius measuring device suitable for use on round work which is being externally ground. Sketch a suitable measuring scale, observing ergonomic principles of design.

5. With the aid of sketches show pneumatic or hydraulic auto sizing devices which are (a) indirect, and (b) direct acting in principle. What are the advantages and disadvantages of each type?

6. Outline a complete, fully automatic sizing and inspection device suitable for use upon a high output, high accuracy machine tool. The whole process must be fully automatic, and self correcting on every feature.

7. Digital slide positioning units are being successfully used for all manner of applications. Choose and describe a machine tool process where you think the use of such an instrument would be economically sound. Outline the principle of operation of the system.

8. A friction roller, in-process measuring device has rollers of 300 mm diameter, and is equipped with angular transducers transmitting 850 pulses/roller revolution. 21 081 roller pulses are counted during 12 revolutions of the workpiece. Calculate the workpiece diameter.

(Ans. 620·03 mm.)

9. Using diagrams, show some form of optical system being used for both indirect and direct auto sizing. What are the advantages and disadvantages of these systems?

Further Reading

1) Bryan G. T. *Control Systems for Technicians.* Hodder and Stoughton Ltd.

2) Charnley C. J. *The Principles and Application of Size Control in Engineering Practice.* H.M.S.O., 1963.

3) Tillen R. 'Controlling Workpiece Accuracy.' *Journal of Institute of Production Engineers,* 1964.

4) Tillen R. and Pearce D. C. K. 'In-process Gauging for Lathes.' *Journal of Institute of Production Engineers,* Oct., 1967.

5) Tipton H. 'In-process Measurement and Control of Workpiece Size.' *Metalworking Production* 110, No. 49.

CHAPTER 10

Special Machining Processes

10.1 ELECTRICAL MACHINING

IN RECENT years the metallurgical side of the engineering industry has developed and introduced new metals, such as the nimonic alloys, and the titanium alloys for example. Some of these are harder, stronger and more temperature-resistant than the older materials. Only grinding, of the older and more conventional metal removing processes, has had any application in the shaping of these new metals. This has led to the development and application of new metal removing processes, which can be classed at this stage as being of a specialized nature. However, some of these special processes which are described in this chapter have been successfully applied in a wide field upon a variety of materials. Some of them are in their early stage of development and have not started to realize their full potential. Also, it is possible that other special machining processes will be invented and used in later years and will find certain applications which are more suited to them than to other machining processes.

One group of the special machining processes can be classed under a general heading of *electrical machining* processes. These are *electrical discharge machining, electrochemical machining* and *electrochemical grinding*, each being described in Sections 10.2, 10.3 and 10.4 respectively.

The electrical machining processes have one thing in common, i.e., they employ an electric current as a means of removing metal. This means that it is not necessary to convert electrical energy into mechanical energy in order to do the same thing, as is done in the conventional processes, such as single point tool machining or grinding for example.

The advantages of electrical machining are that tool forces do not increase as the work metal gets harder, economic metal removal rate does not decrease as the work metal gets harder, and the tool material does not have to be harder than the work metal.

The obvious limitation of electrical machining is that the workpiece metal must be an electrical conductor.

10.2 ELECTRICAL DISCHARGE MACHINING (E.D.M.)

Electrical discharge machining (EDM), or spark erosion, as it is called, depends upon the eroding effect of an electric spark upon two electrodes used to produce the spark; one electrode being the tool and the other electrode being the work. If both electrodes are made of the same material, it has been found that the greater erosion takes place upon the positive electrode. Therefore the work is made positive and the tool negative, hence giving maximum metal removal rate to the work and minimum wear rate to the tool.

The spark is generated by a heavy electrical discharge across a gap between the electrodes which are immersed in a dielectric fluid, such as paraffin, white spirit or transformer oil. The gap has a critical value, (approx. 0·025 mm to 0·075 mm) which must be maintained throughout the operation as work and tool wear. This is done using a reversible servo motor.

The principle of the process is shown in the diagram at Fig 10.1.

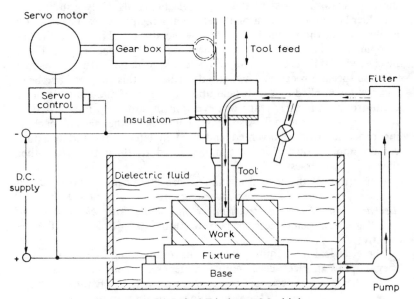

Fig 10.1   Electrical Discharge Machining.

The shaped tool (negative electrode) is fed into the workpiece causing sparking to occur between the closest point of approach between the tool and work. As the tool fully enters the work, sparking occurs equally across the whole area of the tool front face causing the tool form to be reproduced upon the work. Machining, or metal removal, is possible because repetitive sparks of short duration (interval between successive sparks is approx. 0·000 1 second) release their energy in the form of local heat (local temperature approx. 12 000°C), melting a small area of the workpiece and forming a crater upon the work surface. The pumped dielectric fluid, which acts as an insulating medium and provides the correct conditions for efficient spark discharge, also carries away the eroded metal particles. The time interval between the sparks is so short that the heat is unable to conduct into the workpiece.

Let us examine some of the more important features of EDM in detail.

Servo Control
This is a closed loop control system. The tool, held in a chuck, is fed into the work on a vertical slide. This slide has a rack and pinion feed, driven through reduction gearing from a D.C. servo motor. As the current discharges across the gap, the tool and work are eroded causing the gap to increase, and the voltage to drop. This voltage drop is automatically measured hence giving feed back to the control system. The servo control receives a signal causing it to operate the servo motor and feed the tool forward until the gap reaches its critical value again. So sparking takes place again and the whole cycle is repeated. The process is continuous and does not need the services of the operator until the tool reaches a pre-set depth.

Other control systems are used, which instead of sensing a change in gap voltage, sense a change in potential or gap capacitance.

Spark Generating Circuit
Power from an A.C. source is fed to either a D.C. generator, or a rectifier, the D.C. output being fed to the spark generating circuit. The tool and work are connected into this circuit, as is also the tool slide servo-mechanism. The spark generating circuit may be one of two types: (a) relaxation, or (b) pulse-generator. Let us briefly consider each, because they have important differences.

a) *Relaxation circuit.* Most earlier models of spark erosion machines were fitted with a D.C. relaxation type of generator. This supplies current to a condenser, the discharge from which produces the spark. The workpiece alternatively becomes a positive electrode (anode) or a negative electrode (cathode) respectively. On each reversal of polarity the tool is eroded more than the workpiece, hence tool wear is greater with this type of arrangement. As some tool electrodes are of complex shape and expensive to produce, this problem of tool wear is important.

b) *Pulse-generator circuit.* With the development of transistors, later models of spark erosion machines have been available fitted with transistorized pulse-generator circuits in which reverse pulses are eliminated. A closer control can be kept of pulse frequency and also tool wear is greatly reduced. Today, both types of machine are available at varying prices.

The size and frequency (400 to 200 000 Hz) of the electric current governs the spark gap, and the rate of cutting, i.e., roughing or finishing. Controls are provided on the machine in order to allow these electrical parameters to be varied.

Tool Materials

Copper or brass is used extensively as an electrode material for commonplace applications. These materials are perfectly satisfactory as they can easily be shaped to the required profile, but upon production work their wear rate might be uneconomically high. Wear is best defined as the ratio between amounts of material removed on the tool and on the work and is therefore given by:

$$\text{Wear ratio} = \frac{\text{Volume of metal lost upon tool}}{\text{Volume of metal removed from work}}$$

With a brass tool, the wear ratio for the following work materials is as shown:

Work	Wear ratio
Brass	$\frac{1}{2}$
Hardened plain carbon steel	1
Tungsten carbide	3

It has been found that the higher the tool material melting point, the less the tool wear. Hence, the search for suitable tool materials giving less wear has led to the successful use of such materials as tungsten, tungsten copper, tungsten carbide (difficult to shape to a complex profile), copper graphite and graphite.

Using a graphite anode with a pulse generator machine, the wear ratio has been reduced to one-tenth.

Machined Surface

The nature of a spark eroded surface is different to that produced by any other process. It consists of a series of small, saucer shaped depressions produced by the individual sparks, and resembles a matt, shot blasted surface. It is dull and non-reflective. Being non-directional and capable of holding a lubricant well, it is particularly suited to applications of a wear resisting nature.

Surface finishes to the order of $0.2\ \mu\text{m}$ R_a are possible, but it should be remembered when comparing R_a values that a spark eroded surface does not have the peak and valley characteristics of other machined surfaces.

The work surface may be damaged mechanically and thermally. Sometimes a steel workpiece may have a highly carburized surface layer from 0.01 mm to 0.1 mm deep containing up to 4% carbon.

The accuracy of the work produced is highest on a finishing operation when the spark gap and voltage is smallest, and may be as high as 0.005 mm.

Metal removal rates are comparatively low (approx. 80 mm^3/s), although this can be considerably exceeded on the larger modern machines having more efficient pulse-generators. Energy output per unit volume of metal removed per second is comparatively high being to the order of 120 J/mm³.

EDM is widely used for the reproduction of dies for moulding, forging, extrusion and press tools, in difficult materials like alloy steels and tungsten carbide. Dies can be machined in the hardened state eliminating the risk of distortion. Cutting forces are small, so very fine holes can be accurately produced. In die manufacture, the punch has often successfully been used as the electrode for machining. For production work the process has been used for special applications where the oil retention properties of the work surface are all important.

10.3 ELECTROCHEMICAL MACHINING (E.C.M.)

Electrochemical machining (E.C.M.) is quite different from EDM in principle. Again, this is an instance of a process being developed to cope with new hard, tough materials developed by the metallurgists, in this case in the rocket and aircraft industry. The process makes use of the principle of electrolysis in order to dissolve away the workpiece material, and the work material must be an electrical conductor. In effect, the process is the reverse of electroplating. Let us therefore examine the principle of operation of electroplating before considering ECM.

Electroplating

Electroplating is based directly upon Faraday's Laws of Electrolysis which may be summarized thus:

'The weight of substance produced during electrolysis is directly proportional to the current which passes, the length of time of the electrolysis process, and the equivalent weight of the material which is deposited.' The process depends upon the presence of ions in the electroplating liquid solution. Ions are electrically charged atoms which facilitate the transport of electricity through the electrolyte. (The liquid solution used in electroplating is called the *electrolyte*.) Ions may be positively charged, such as metal ions for example, or may be negatively charged, such as sulphate or chloride ions for example. Let us consider copper electroplating which is the most common, this being represented in diagrammatic form at Fig 10.2.

The object to be plated is the negative electrode, called a *cathode*, and may be steel, say. The positive electrode is called an *anode* and will be made of copper. The liquid electrolyte is a solution of copper sulphate ($CuSO_4$). When the current is passed into the electrodes, the electric potential between the plates dissociates the $CuSO_4$ into positive copper ions and negative SO_4 ions. The Cu ions are attracted to the cathode and are deposited as metal.

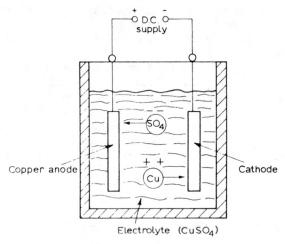

Fig 10.2 Electroplating Process.

The SO_4 ions are attracted to the anode where they attack the copper, hence causing more $CuSO_4$ to be formed. This replaces that removed from the solution by the electrolytic process. The sulphate ions play no direct part in the electrochemical process just described, which may be regarded as the transferring of copper in the ionic state from an anode to the cathode. The anode loses metal to the cathode.

Principle of ECM
In this case the reverse of electroplating is applied, so the work-piece becomes the anode, and the tool the cathode. Therefore, the work loses metal, but before it can be plated on to the tool, the dissolved metal is pumped away in the flowing electrolyte.

A diagram of the process is shown at Fig 10.3.

Figure 10.3 shows that, unlike E.D.M., it is not necessary for the work to be submerged in the liquid solution, but the electrolyte is pumped around at high speed. There is no tool to work contact, but again unlike E.D.M., there is no tool wear, giving very long tool life.

Let us examine some of the more important features of E.C.M. in detail.

Tool Feed
As the tool does not wear away during machining, it is not necessary to have a servo control system in order to maintain a constant gap between tool and work. The tool must be fed into the work at a constant rate, this feed rate depending upon how fast it is required to remove the metal. This in turn depends upon the workpiece material and current density. The relationship between these parameters is known, and the metal removal rate for a given set of conditions can be predicted quite accurately. Hence, an appropriate constant feed rate can be selected.

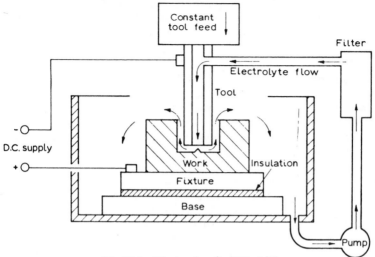

Fig 10.3 Electrochemical Machining.

A gap of 0·01 mm is maintained between the tool face and the work. At this point where tool face and work are closest, the electrical resistance is lowest and hence the current is highest. Therefore the work-metal is dissolved most rapidly in this region.

The movement of the tool slide can be controlled by a hydraulic cylinder giving an infinite range of feed rates.

Power Supply
A relatively simple circuit is used in which the work is connected to the positive terminal of a D.C. supply, and the tool connected to the negative terminal. The difficulty lies in the fact that the D.C. supply must be low voltage (2 to 20 volts), but very high current (up to 10 000 amps). The electrical generating equipment can represent the major part of the fixed costs of an E.C.M. installation.

Electrolyte
The electrolyte serves two purposes. Firstly, it is necessary in order for the electrolytic process to work, as described under electroplating earlier. Secondly, it removes the heat generated in the cutting zone due to the flow of high current. The electrolyte must be pumped around at speeds of at least 30 m/s in order to constantly replenish the solution, and also to avoid overheating which would cause a disturbance in the current flow. The electrolyte must never be allowed to reach boiling point.

In principle, any electrolyte will be satisfactory which is chemically active enough to cause efficient metal removal. In practice, however, electrolytes which are very corrosive are not used. In any event stainless steel parts are used for pipes, etc., which come into contact with the liquid. Salt solutions such as brine, or a 10% solution of sodium chloride are used. Also solutions of sodium nitrate or acids have been used. There is a problem in finding pumps capable of delivering hot acid or salt solutions without pulsation at the required volume and pressure.

Tool Design

Any material which is a good conductor is satisfactory as a tool material, because no wear takes place. However, the tool must be strong enough to withstand the hydrostatic forces which may be considerable. These forces are caused by the electrolyte being forced at high speed through the gap between tool and work.

Unlike E.D.M., the tool shape is not always precisely reproduced upon the work, and with complex shapes this can be a difficult problem. The current density must be uniform over the whole cutting surface in order to give exact workpiece reproduction. In practice, current density may vary due to a variation in electrolyte conductivity for example, which in turn is brought about by changes in temperature across the area of the surface being machined. Other factors to do with the current flow through the electrolyte can lead to a correction of tool shape being necessary.

Machined Surface

As the work temperature cannot be raised above the boiling point of the electrolyte, no thermal or mechanical damage is caused to the machined surface. Furthermore the surface is stress free.

Surface finish values as low as $0.1 \, \mu$m R_a are possible. The process does not require this to be achieved in roughing and finishing stages, but fine finishes can be obtained in one cut. By happy chance the highest possibly current density gives both the highest machining rate and the best finish. Therefore, unlike conventional machining, the faster the cut the better the finish. No burrs are produced.

The accuracy of the work produced is high, and at least equal to E.D.M., i.e., to the order of 0.005 mm.

Metal removal rates are high, up to $550 \, \text{mm}^3/\text{s}$, and depend only upon the chemical composition of the metal being unaffected by the hardness and toughness of the work material. The current consumption per unit volume of metal removed per second to achieve these fast rates is very high, and at its present stage of development E.C.M. is an expensive process and is probably only economic working upon metals harder than 400 BHN. E.C.M. has been successfully applied to the machining of complex shapes in very hard materials, especially those to be used in high temperature applications such as jet engine parts. It is true to say that this process is as yet only at the early stages of its development, and in later years will find many more applications at present carried out by other means.

10.4 ELECTROCHEMICAL GRINDING (E.C.G.)

E.C.M. has been adapted in several ingenious ways and can simulate most of the conventional machining processes. One now finds different machines available which can carry out electrochemical milling, electrochemical shaping, electrochemical honing and electrochemical grinding for example.

The latter is becoming an important process in its own right, now being widely used for the grinding of tungsten carbide tool tips. The advantage of E.C.G. over conventional grinding is that higher metal removal rates are possible, particularly upon hard materials without inducing grinding cracks. Also the process is more economical in the use of abrasives, which is particularly important if the abrasive is expensive, like diamond grit for example.

Figure 10.4 shows the principle of operation.

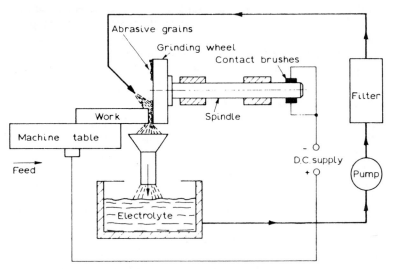

Fig 10.4 Electrochemical Grinding.

It can be seen that the process is similar to conventional grinding in that an abrasive grinding wheel is used, and the work is fed against the rotating wheel. In fact, 10% of the work metal is removed by abrasive cutting, and 90% by electrolytic action. The wheel and its spindle are insulated from the rest of the machine, and are connected to the negative terminal of a low voltage D.C. supply. The workpiece is connected to the positive terminal. Hence, current flows between the work and wheel.

The work (anode) and wheel (cathode) do not make direct contact with each other because they are kept apart by the insulating abrasive particles which protrude from the face of the grinding wheel. A constant gap of 0·025 mm is maintained into which a stream of electrolyte is directed. The electrolyte is carried past the work surface at high speed by rotary action of the grinding wheel.

The grinding wheels used are of conventional shape and structure. Metal bond, diamond grit wheels are used for grinding tungsten carbide tips. Carbon bond wheels are used upon the hard alloy steels, such as the stainless steels. Wheel wear is negligible because the greatest part of the cutting action is electrolytic, and little dressing is necessary, hence wheel usage is very economical. Inefficient conventional grinding leads to work distortion and surface cracks, but there is less danger of these faults occurring in E.C.G. because temperature rise and cutting forces are small.

Accuracy is reasonably high (0·01 mm) but depends upon the abrasive part of the process, and is controlled by the nature of the abrasive surface of the wheel. Therefore, the choice of the wheel grit size, etc., is still important if the desired results are to be achieved. A surface finish of 0·1 μm R_a is possible by the correct choice of the many parameters involved.

As with all grinding processes, metal removal rates are comparatively low being to the order of 15 mm^3/s, and power consumption is high.

10.5 ULTRASONIC MACHINING (U.S.M.)

This is a very specialized metal removing process being mainly confined to the drilling of circular or non-circular holes in exceptionally brittle materials. It has found a place for itself in shaping the very brittle, electrically non-conducting materials such as glass and ceramics, and the semi-precious stones such as sapphires. It has been used for fine drilling holes as small as 0·01 mm diameter in these materials. The equipment is small in scale, and the application very simple indeed. The process is most innocuous to observe with apparently very little happening; the finished results however, on apparently unworkable materials, being most efficient.

The principle of operation is that of using a tool vibrating at ultrasonic speeds, frequencies to the order of 20 000 Hz being used. (Note that in Section 2.5 it was stated that the human ear is unable to detect frequencies above 15 000 Hz.) An amplitude of 0·05 mm is used. In operation, the tool cannot be seen to be reciprocating or vibrating, but when touched this can be sensed.

Figure 10.5 shows a diagram of the process.

The tool vibrates against the work surface while an abrasive slurry is fed between the tool cutting surface and the work. The loose abrasive in the slurry is hammered into the work surface, the material being removed by the chipping action of each abrasive particle. Feed rates of the tool through the work are low being up to a maximum of 0·1 mm/s. The abrasive used is usually boron carbide in a very fine grit size, suspended in water to give the slurry. The water also acts as a coolant. Silicon carbide in paraffin has also been used.

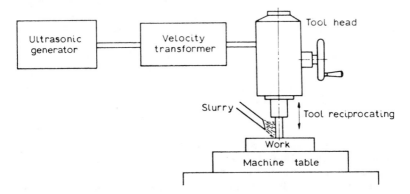

Fig 10.5 Ultrasonic Machining.

The tool material can be metal which should be hard and tough, but not brittle or the tool cutting face will be chipped. The tool can be of any profile which will be reproduced in the work. The accuracy of the finished profile depends upon the tool accuracy, but may be as high as 0·003 mm, but surface finish quality is not high, and may not be better than 0·4 μm R_a.

An ultrasonic generator is required to provide a means of power. This will be used through a velocity transformer in order to give the required amplitude which will make the process successful. Generators of 100 watts output are suitable for small work, but larger units are being built up to 2 000 watts output which are capable of driving tools having several cm² of cutting surface area. The rate of metal removal is low (3 mm³/s maximum), and the power consumption is high, but when used for the special applications mentioned, economic considerations will be secondary. However, attempts are being made to extend the use of the principle to softer materials and this has been done by combining ultrasonic and electrochemical machining in the ratio 10 – 90. This is a more economic proposition because the greater accuracy of the ultrasonic process is combined with the greater metal removal rate of electrochemical machining.

Much research is being carried out with a view to improving and extending the new machining processes. Ultrasonic machines are available with new features such as diamond impregnated tools for example. Also on some machines the tool rotates at up to 1 500 rev/min as well as reciprocating. This gives increased accuracy of drilling, and holes can be drilled up to several centimetres deep in ceramic material for example, with great accuracy.

10.6 ELECTRON BEAM MACHINING (E.B.M.)

The last three sections in this chapter all refer to more exotic processes which are in their early stages of development as viable commercial machining methods. All three processes are said to be of outstanding value with apparently great potential. Let us consider E.B.M. first.

The process uses a high energy beam of electrons which can be focused magnetically upon a very small area, say a spot of 0·003 mm diameter. The kinetic energy of the beam (travelling at about half the velocity of light, i.e., 160 000 km/s), is converted into heat energy upon striking the workpiece, and raises its temperature locally in that very small area to above boiling point. Hence, the material is vaporized and very fine holes for example could be drilled in difficult materials, such as sapphire. (Note. An electron is the lightest known particle. It is a constituent of all atoms around whose nuclei they revolve in orbits. The force which holds an electron in orbit is electrostatic, and it carries a negative electric charge. Hydrogen, to take an example, has one electron moving around the nucleus of the atom, has the atomic No. 1, and is the first element in the atomic table. An iron atom has 26 electrons and iron has the atomic No. 26.)

Figure 10.6 shows the principle of operation of E.B.M.

It can be seen from Fig 10.6 that an electron gun is used to form the electron beam, and the beam focusing and deflecting is controlled in a similar way to that used in a television tube. A cathode ray tube

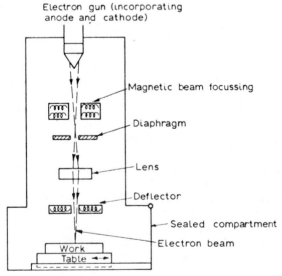

Fig 10.6   Electron Beam Machining.

used for T.V. transmits electron beams at high speeds to build up the lines of the picture with spots of light, left to right along a line, line by line. Electron beams used for machining are far more powerful than those transmitted for T.V., and a supply of 10 to 100 W is used. The process must be carried out in a vacuum, and means must be provided to place the workpiece in a sealed compartment.

An optical viewing system is required in order to allow the operator to see the work, in order to position the beam. The deflecting coils allow the beam to have a range of movement of 12 mm. The electron beam can be controlled very accurately, and machining tolerances of 0·005 mm are possible. Power consumption is exceptionally high, but applications at the moment are very specialized, and the normal machining parameters have little significance. The beam is transmitted at 20 000 Hz, and the thermal effects can be controlled in such a way that only the surface layers of the work material are affected.

The process is not suitable for sinking deep holes where side walls must be parallel. It can be used for drilling or cutting, and holes or slots of less than 0·002 mm can be produced in thin materials. It is particularly suitable when used upon materials having a high melting point combined with low thermal conductivity. Automatic production machines are available having more than one beam, with production rates of several 100's/hour. These are used for example, in drilling synthetic jewels in the watch industry.

10.7 LASER BEAM MACHINING (L.B.M.)

L.B.M. is similar to E.B.M. only in that both processes utilize a high energy beam which can be focused upon a small area. Therefore, their applications are similar and the processes are comparable in that sense. In fact, the nature of the beams is entirely different. A laser transmits amplified light, and the word laser stands for 'light amplification by stimulated emission of radiation'. The term is complicated and the theory more so. However, as throughout this book, we will consider the theory not for its own sake, but only in sufficient depth to appreciate the economics and scope of the process.

The two types of *laser* which are of significance in materials processing are the *solid-state* and *gas* respectively.

Solid state laser

This utilizes a solid fluorescent material capable of radiating and amplifying light. Materials used to date are ruby (crystalline aluminium oxide or sapphire), YAG (yttrium-aluminium-garnet), or Nd-glass (neodymium-in-glass). The material is fabricated into rods with optically finished ends.

The laser material needs a source of energy called a pump. This may be an optical flash tube which excites the fluorescent atoms in such a way that light is given off by stimulated emission. (Compare this to the spontaneous emission of a normal lamp.) The short pulse of intense light emitted will have an energy of several joules lasting for a minute fraction of a second, the input power being 2 000 to 10 000 watts. For machining, powerful short pulses of say 100 joules energy are required, and this can be done using a special technique known as Q-switching. The laser light is monochromatic and can be focused optically on to very small spots of less than 0.002 mm dia. hence melting and vaporizing the work material through electro-magnetic radiation.

Gas laser

Instead of a rod of solid material a gas is used such as argon, or more commonly CO_2 (carbon dioxide). Figure 10.7 shows the principle of operation.

The stimulation lamp shown in Fig 10.7 is a tungsten-halogen (see Section 2.7) or krypton-arc flash lamp with a power source of 250 to 1 000 watts. This is located near the gas medium (or solid crystal rod in the case of a solid-state laser). The light from the lamp is reflected and

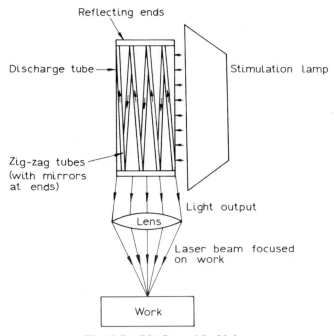

Fig 10.7 CO_2 Laser Machining.

focused onto the gas by mirrors located inside the discharge tube. As described earlier this 'optical pumping' process excites the atoms in the laser medium (in this case, gas) and raises them to a high energy level. This results in the emission of a highly amplified light beam. In turn the light beam from the laser is focused onto the workpiece.

A laser then, is a device which generates, amplifies and emits a concentrated beam of light energy, which is several times brighter than the light emitted by the sun. It has now found a permanent place in the field of micro-machining, particular in the drilling of fine holes in special materials. In some cases it has proved to be the only practical way to drill certain products, Solid-state lasers are now used extensively in the manufacture of micro-electronic circuits, and CO_2 lasers have been found especially useful for processing plastic materials.

Finally, as E.B.M. and L.B.M. are so similar in application it may be useful to make some comparison between them:

a) The greatest single advantage of laser over electron beam is that the laser needs no vacuum.

b) Power densities (the output in watts which can be concentrated upon an area, usually expressed in units of W/mm^2) are greater for lasers than electron beam.

c) The laser beam cannot be deflected electrically, so that the movement of the beam with respect to the workpiece must be carried out mechanically. Therefore the laser beam cannot be controlled as accurately, and machining tolerances are less, being to the order of 0.01 mm.

10.8 PLASMA JET MACHINING (P.J.M.)

All gases burning at high temperatures are ionised gases, and the use of oxygen-acetylene and other similar gases for metal cutting is known to every engineer. Plasma torches are similar, but the chemical energy of the flame is increased by the addition of electrical means, hence raising the flame temperature. Compare plasma temperatures of 16 500° C with oxyacetylene temperatures of 3 000° C.

In a plasma gun, an inert gas such as argon is passed through a small chamber in which an arc is maintained. Electrons flow from the cathode to the anode, this flow ionizing the gas. The gas molecules then disassociate causing large amounts of thermal energy to be liberated. The ionized gases and arc are forced through the nozzle of the plasma gun. (Note. The smallest particle of gas which can be produced and still be gas is called a molecule. Hence, a molecule can be defined as a group of atoms.)

In principle there are two main types of plasma arc systems:

a) The transferred arc.

b) The non-transferred arc.

These are shown at Fig 10.8(a) and (b) respectively.

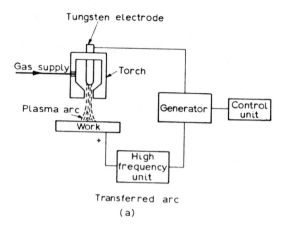

Transferred arc

(a)

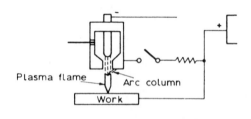

Non-transferred arc

(b)

Fig 10.8   Plasma Arc Machining.

Figure 10.8 shows the fundamental differences in the two systems. At (a) it can be seen that the arc is struck between the work and the torch electrode. In effect the work is the anode and the electrode is the cathode. In this case more of the electrical energy is transferred to the work, hence giving the work more heat.

At (b) the arc is struck between two parts of the torch. In effect the torch case is the anode and the torch electrode is the cathode for the purposes of striking the arc. The plasma jet or flame feeds the heat to the work. In both cases a copper alloy nozzle is used on the torch, both the nozzle and the electrode being water cooled.

Plasma systems are used for metal cutting and gouging in difficult materials such as stainless steel. In addition they can cut plain carbon steel plate four times faster than can the ordinary flame cutting torches. On the machining side, plasma has been used for rough turning of very difficult materials. The torch is held near the rotating work, tangentially to and below it, at an angle of 30° to the horizontal. This is a roughing operation to an accuracy of about 1·5 mm with a corresponding surface finish. Energy output per unit volume of metal removed per second is to the order of 140 J/mm^3.

Exercises 10

1. What is meant by electrical machining?
Describe and compare the different variations of this process.

2. Clearly show the principle of (a) E.D.M. and (b) E.C.M. Compare the economics of metal removal of each, and their different applications.

3. What is the difference between E.C.G. and orthodox grinding?
What are the advantages of the former that sometimes make it an attractive alternative to the latter?

4. With the aid of sketches compare and contrast L.B.M. and E.B.M.

5. What is the principle of the plasma arc system and how does it differ from orthodox flame torch systems? Has it any applications in the metal removing field?

Further Reading

1) Boden P. J. and Brook P. A. 'Principles of E.C.M.' *Journal of the Institute of Production Engineers*, Sept. 1969.

2) Jukes A. 'Machining by Spark Erosion.' *Journal of the Institute of Mechanical Engineers*, Oct. 1968.

3) Bock F. 'The Plasma Torch in Metal Cutting.' *Metalworking Production*, May 1964.

4) 'Electron Beam Processing.' *Aircraft Production*, Aug. 1960.

5) Kennedy D. C. and Grieve R. J. 'Ultrasonic Machining – a review.' *Journal of the Institute of Production Engineers*, Sept. 1975.

6) Noble C. F. and Sfantsikopoulos M. M. 'Electrochemical surface Grinding.' *Journal of the Institute of Production Engineers*, May 1976.

7) Holland M. B. 'Electrical Machining – whatever happened to it?' *Journal of the Institute of Mechanical Engineers*, May 1979.

8) Crookall J. and Shaw T. 'Why E.D.M. is expanding from the tool-room onto the shopfloor.' *Journal of the Institute of Production Engineers,* Jan. 1980.

9) Webster J. M. 'The Application of CO_2 lasers to Manufacturing Processes.' *Journal of the Institute of Production Engineers,* July/August 1976.

10) 'Putting the Laser to use.' *Journal of the Institute of Production Engineers,* May 1978.

CHAPTER 11

Production of Plastics

11.1 POLYMERS

THERE IS no generally accepted definition of a plastic, but it can be regarded as a material containing a synthetic high polymer as the major constituent. Polymers are materials of high molecular weight formed by joining together many (poly) small molecules. In some plastics the polymer may be the only constituent, but more usually a variety of other substances are incorporated such as fillers, pigments, antioxidants, plasticizers, etc. These additives will enhance specific properties and/or facilitate processing. Almost all commercial plastics at present are based upon organic polymers. The word plastics is closely associated in many minds with cheap, brightly coloured products of limited life. The time has probably come to adapt a new name more worthy of the exciting range of polymer materials with many different properties which are available to engineers.

From the production point of view the main advantage of plastic materials is their relatively low melting point and their ability to flow into a mould. Generally there is only one production operation required to convert the chemically manufactured plastic into a finished article. Contrast this with metal where generally many operations are required to convert the raw material into its final shape.

It is convenient to divide plastics into two groups depending upon the geometrical form of the polymer molecule. The first group are the linear polymers which are the basis of *thermoplastics*; the second group the network polymers which are the basis of the *thermosetting plastics*.

11.2 THERMOPLASTICS

In linear polymers the molecules are synthesized in the shape of long threads. They are called *thermoplastics* because they soften upon heating and harden upon cooling in a reversible fashion. In the melted state they are rubber-like liquids, and in the hard state they are glassy and brittle (e.g. Perspex) or partially crystalline (e.g. Nylon). The latter type decrease in volume upon solidification by several per cent which must be taken into account during processing.

Properties of some of the thermoplastics are given below in the table shown at Fig 11.1.

Polymer	Tensile Strength	Compressive Strength	Machining Properties	Chemical Resistance
Nylon	Excellent	Good	Excellent	Good
P.T.F.E. (Poly-tetra-fluoro-ethylene)	Fair	Good	Excellent	Outstanding
Polypropylene	Fair	Fair	Excellent	Excellent
Polystyrene	Excellent	Good	Fair	Fair
P.V.C. (rigid) (Poly-vinyl-chloride)	Excellent	Good	Excellent	Good
P.V.C. (flexible)	Fair	Poor	Poor	Good

Tensile Strength	Excellent	55 N/mm²
	Poor	21 N/mm²
Compressive Strength	Excellent	210 N/mm²
	Poor	35 N/mm²

Fig 11.1   Table of Properties of Thermoplastics.

Thermoplastics can be processed to their final shape by moulding or extruding. Extruding is often used as an intermediate process to be followed by vacuum forming or machining for example. We will consider each of these processes in turn.

11.3 MOULDING OF THERMOPLASTICS

Injection Moulding of Thermoplastics

Injection moulding is the most important of the plastic moulding processes. Moulding machines range in size from an injection capacity of 12 000 mm³ to $2 \cdot 2 \times 10^6$ mm³, and locking forces applied to the mould from $0 \cdot 1$ MN to 8 MN and occasionally greater, usually applied by hydraulic means. This process can be compared to the pressure die casting of metals, and Fig 11.2 shows the principle of operation of the process.

The cycle of operations for the machine are:

a) Feed a metered amount of polymer powder into the cold end of the injection cylinder.

b) The injection ram forces the powder forward into the heating section of the cylinder where its temperature is raised to the order of 150° C – 300° C (e.g., P.V.C., 175° C moulding temperature; polypropylene, 260° C moulding temperature). Moulding pressure is built up the order of 100 N/mm² – 150 N/mm².

c) The pressure build up forces the previous charge to inject into the mould cavity and is maintained after the mould is filled.

d) The ram returns and the polymer in the water cooled mould solidifies hence sealing the mould cavity. The mould opens at the parting line and the component is ejected. The mould closes and the cycle is repeated.

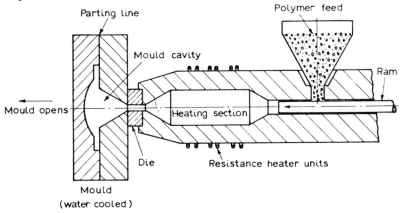

Fig 11.2 Ram Injection Moulding Machine.

As with pressure die casting the mould may have several impressions hence producing several components for one machine cycle. Although Fig 11.2 illustrates the principle of injection moulding, ram or plunger type machines can now be considered old fashioned. Modern machines make use of a screw preplasticiser unit instead of a ram. See Fig 11.3.

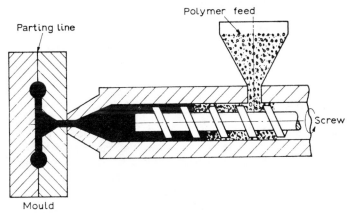

Fig 11.3 Screw Plasticiser Injection Moulding Machine.

This serves the function of both heating the polymer by internal shear and feeding it to the mould. A proportion of the heating is still done by resistance heater bands upon the barrel as in ram machines. The advantage of heating by internal shear is that it is self regulating in that as polymer temperature rises, the viscosity falls. Hence less heat is generated for the same amount of shear. The screw plasticiser is the best method of heating thermoplastics because overheating is avoided, and the plastic charge remains in the heating chamber for a much shorter time because throughput is continuous.

The advantages of the process are similar to the pressure die casting of metals viz.,

i) Wide range of shapes and plastic materials can be moulded.

ii) Automatic or semi-automatic (using unskilled labour) reproduction of large numbers of accurate components. Accuracies of $\pm$ 0·025 mm. can be maintained on some of the easily moulded polymers.

iii) Manufacture is fast. The cycle time of an injection moulding machine is only a few seconds. Blow moulding is slower.

Blow Moulding

This process was inspired by the glass blowing process used for the mass production of glass bottles and containers. The process commences with the extrusion of the blank called a *parison* which is transferred to a mould which is then closed. The parison is blown to its finished shape inside the mould. Much vin ordinaire wine is now bottled in France in P.V.C. blow moulded bottles used once only before disposal.

Casting

The gravity casting of thermoplastics in moulds is carried out very little compared to the gravity die casting of metals which is the comparable process. This is because the viscosity of thermoplastics cannot be reduced sufficiently by raising the temperature to allow them to flow easily as metals do under gravity into a metal die. One exception is nylon monomer mixed with suitable agents which give a free flowing liquid. Many large nylon parts such as gears have been successfully gravity die cast, this process being suitable for components of mass from 0·5 kg to 50 kg.

11.4 EXTRUSION PROCESS

This often serves the function of producing the plastic in an intermediate form for subsequent reprocessing, as well as being used to produce a polymer in its final form. Therefore it can be said to be the most important of the plastic processes and consequently has received a lot of attention. The actual extruding part of the process is similar to that used for metals in that the material flows through a die orifice of the required shape under pressure. After leaving the die the polymer swells appreciably, the amount of swelling depending upon the shear rate (due to the screw) and the molecular weight distribution of the polymer. If the die is of non-uniform cross section the pressure upon the die walls will not be uniform and it has been found that die swell then will not be uniform. Therefore it can be seen that designing and making dies for varying polymers is an uncertain business, and a lot of trade 'know-how' and trial and error is required.

The earlier stages of the extrusion process when the material is brought to a state suitable for extrusion is not similar to that used for metals as the properties of plastics are so different. As with injection moulding, ram type machines have been used in the past, and some are still used. However they have some disadvantages. The ram pressure will usually be insufficient to plasticize the polymer hence a heating chamber will be necessary or the material will have to be pre-processed in order to soften it. Also the ram has an idle return stroke which means the process is not continuous and is consequently slower. Finally the die orifice may have to be cleaned in between each working stroke. It can be said that today ram type extrusion machines are confined to the field of wet extrusion, i.e., where the polymers are plasticized and softened by pre-processing using a solvent.

Screw Extrusion Machines
A screw is the best means of feeding and plasticizing a polymer for the reasons given under Section 11.3 on injection moulding machines. The principle of operation is shown in Fig 11.4.

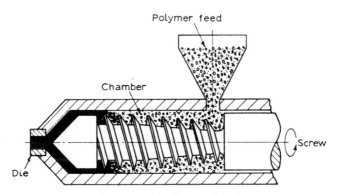

Fig 11.4 Screw Extruding Machine.

The screw has a special thread form to suit the material being extruded and is water cooled. Some heating may be provided to the chamber in addition to the heat provided by the action of the screw.

The operation of the machine is continuous and consists of:

a) Polymer granules or particles are fed into the machine hopper.

b) The polymer feeds from the hopper through an inlet port to the screw. The screw rotation imparts both an axial motion and a rotary motion to the material causing compression and shear respectively. Therefore the material does not in fact flow through the die in lines parallel to the die axis (as in metal extrusion) but follows a spiral path. This is a further complication affecting die design.

c) As the polymer feeds through the chamber pressure builds up due to the restricting effect of the die opening. The polymer mass is worked and sheared by the screw, the frictional effects raising the temperature until the whole is in a plastic state and ready for extrusion.

d) The plastic is extruded through the die, finally cooling and hardening.

Some extruding machines may be melt extruders rather than plasticizing extruders as just described. The difference is that the material is delivered to the machine hopper already melted. Therefore the machine merely acts as a constant delivery pump, and the polymer is pumped through the die orifice.

Colouring of plastics can be carried out by adding dry colour to the screw extruder (or the screw plasticizer of an injection moulding machine). Many thermoplastic products are extruded, an example being P.V.C. domestic water pipes which may be buried in the ground without deterioration for indefinitely long periods of time.

11.5 SHEET FORMING PROCESS

This is a plastics process which is comparable in the class of work produced with the stretch forming of metal sheet. The plastic sheet is hot formed, vacuum equipment most usually being used. Vacuum forming enables the polymer to be converted into shape in one operation which would not be possible with metal.

The process starts with a calendered sheet of thermoplastic which is heated until the material is in a rubbery condition. It is then draped and clamped over the mould in such a way that the air between the sheet and the mould can be evacuated. A vacuum of increasing intensity is applied through the male former, and the sheet of plastic under the influence of atmospheric pressure on the outside is forced to conform closely to the contours of the mould. As the sheet is drawn tightly to the mould it stretches, eventually cooling and solidifying.

The principle of operation is shown in Fig 11.5.

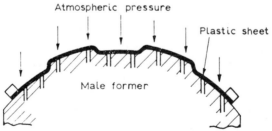

Fig 11.5 Principle of Vacuum Forming Process.

The sheet can be moulded to any shape and there are many variations of the process. For example female formers can be used, or reverse drawing can sometimes be carried out as with sheet metal. The choice will depend upon the shape of the part and the type of polymer to be used. The heating which will only take one or two minutes can be done from one side only, but preferably and more quickly on both sides of the sheet (sandwich heating). As heating proceeds the polymer sheet begins to expand and sag. Contraction follows as the stresses set up by the extrusion process are released. The sheet then may wrinkle and pucker. These wrinkles disappear as heating proceeds and the material becomes clear and translucent, and this indicates that the sheet is ready for forming. This only happens with natural materials which are not coloured by pigmentation. Due to the poor thermal conductivity of polymers, the sheet cools slowly when the heaters are withdrawn and this allows time for manipulation.

The amount of stretching which can be imposed upon a hot plastic sheet depends upon the stretching temperature and the rate of stretch. At a constant stretching speed a higher temperature will give higher elongation up to a certain critical temperature. Above this temperature elongation is reduced. See Fig 11.6. The slower the rate of stretch the higher will be the elongation, but of course the sheet has more time to cool. Hence the benefits of a higher temperature level are lost. In practice, as with many engineering applications, a compromise is reached between temperature and stretching rate. A polymer with a wide temperature range at the required amount of elongation is ideal, such as plastic No. 1 in Fig. 11.6.

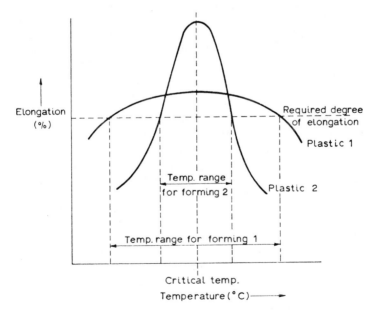

Fig 11.6 Temperature Range and Critical Temperature for two Thermoplastics

A polymer which behaves like plastic No. 1 (Fig 11.6) is P.V.C. for example, and a polymer which behaves like plastic No. 2 (Fig 11.6) is polystyrene.

Cold Sheet Forming

There are certain technical difficulties due mainly to the marked creep effects of plastics at normal temperatures which make cold forming of plastic sheet an impracticable proposition. However, it has been found that by stretching certain plastics in the direction of extrusion, and at the same time stretching it at right angles to that direction, a material is obtained which can be processed upon conventional metal forming tools. Hence, it may be found in the future that biaxially orientated stretched plastic sheet can readily be drawn, deep drawn, and press brake formed and flanged.

11.6 MACHINING OF THERMOPLASTICS

For small production runs it will often be necessary to machine the plastic work to shape from extruded rod or sheet rather than to mould to shape using expensive dies. Practically all plastic materials can be machined satisfactorily. In general they should be treated as the soft metals (particularly aluminium or brass) in that a high speed and fine feed is required. High speed steel can be used as a cutting tool material, but for long runs tungsten carbide is better particularly on the filled materials, as it will enable a sharp edge to be retained longer on the tool cutting face. It is important to maintain a sharp cutting edge with ample clearance otherwise overheating will occur with consequent poor finish. Overheating must be avoided at all costs otherwise the polymer may well melt or burn. Coolants are usually necessary, water or a soluble oil solution being adequate. Nylon for example will often machine better using a light cutting oil.

Woodwork machines are often more useful than metalworking machines for machining plastic. For example, sheet or rod is best sawn to size using a band sawing machine rather than a reciprocating hacksawing machine. Some polymers produce dust or powder when machined, and extraction equipment must be provided in order to keep the air clean. With this type of plastic the best coolant will be compressed air which will also keep the cutting zone free from the fine swarf. Metalworking machines will often need to be adapted if they are to be used for an extended run of plastic machining. Firstly, the speed range will often not be nearly high enough, and the driving arrangement will need modifying (unfortunately the bearing arrangement may then be found to be unsatisfactory for high speeds although more than adequate for the very small cutting forces involved). Secondly, the work holding arrangement and capacity may be found unsatisfactory as plastic materials are so easy and light to handle. Thermoplastics are in general much more flexible than metals, and difficulty may be experienced in the work bending away from the cutting tool. It will then be found necessary to use steadying and supporting devices. Thermoplastics are too soft to be machined satisfactorily by the abrasive processes such as grinding. A high and adequate degree of finish can be obtained using high speed steel or tungsten carbide cutting tools.

The following table shown at Fig 11.7 gives a general guide for cutting tool angles, speeds and feeds for the common thermoplastics.

| Material | Rake | Clearance | Turning | | Milling | | Drilling | |
			Cutting Speed (m/s)	Feed (mm/rev)	Cutting Speed (m/s)	Feed (mm/s)	Cutting Speed (m/s)	Feed (mm/rev)
Nylon	0 to −10°	20° to 30°	2·5 to 5	0·1 to 0·25	5	up to 4	2 to 5	0·1 to 0·38
P.T.F.E.	0 to −5°	20° to 30°	1 to 2·5	0·05 to 0·25	5	up to 4	1·25 to 5	0·1 to 0·38
Polystyrene	0 to −5°	20° to 30°	1·5 to 5	0·05 to 0·25	5	up to 4	0·5 to 10	0·1 to 0·38
P.V.C. (rigid)	0 to −10°	20° to 30°	1·5 to 5	0·25 to 0·75	5	up to 4	2·5 to 30	0·05 to 0·13

Fig 11.7 Table of cutting tool angles, speeds and feeds for Machining Thermoplastics.

Some of the more important machining processes for thermoplastics will be considered in turn.

Turning

The design of the tools is the most important feature and with the great variety of thermoplastic materials the best combination of tool shape and angles may only be found by experimentation. In general negative rake and large clearance angles are necessary, and it is most important that the tool is kept sharp. Wherever possible use roller steady box tools with a tangential tool to prevent flexing of the work. Where very fine powdery swarf can clog the rollers, use vee steady box tools as an alternative. Some swarf, such as nylon, will be found to be long, and extremely tough. It is not easily broken from the work, and chip breakers do not work.

Whether a heavy depth of cut or a light depth of cut is taken it is usually possible to achieve a high degree of finish. Plunge cutting techniques such as forming or parting off are possible. In the case of forming it is recommended that the length of the cutting edge should not be wider than the minimum work diameter. Normal parting off blades can be used, and they will be found to perform best if the cutting edge is parallel to the work axis; ample side clearance and negative rake being provided.

A large volume of coolant should be provided to prevent melting or burning, but a high pressure of coolant delivery is not required.

Milling

No special techniques are required except those already generally stated. Cutting speeds should never fall below 5 m/s, and surface finish will improve with finer feeds irrespective of the depth of cut used. A sharp cutter will give best results. Normal cutters can be used for horizontal milling and climb milling is recommended to avoid burning. Vertical milling is ideal using end mills or fly cutters. High speed vertical milling or the woodworking process of routing is a very successful technique applied to certain of the thermoplastics.

The work often has to be carefully clamped and supported to prevent flexing away from the cutter. Vacuum chucks may be found ideal for some materials.

Drilling

It is very easy when drilling thermoplastics to overheat the material, and feeds should not be too coarse. Ample quantities of coolant are necessary. On many materials the hole closes in after drilling and it may be necessary to drill oversize. It is imperative to allow the swarf to clear quickly, otherwise a swarf build-up in the drill flutes will cause over-heating with consequent spoiling of the work. Two things will help to avoid this. Firstly, a drill having polished flutes with a slow helix angle will facilitate swarf discharge. Secondly, drilling should be done by hand feeding using the *woodpecker* technique, i.e., the drill should be withdrawn after each small increment of feed to clear the swarf.

Depending upon the polymer being drilled the point angle can vary from 80° to 130°. Mostly the metalworking drill angle of 118° will do for general purposes. The lip clearance angle can vary from 8° to 16°. Again the work should not be allowed to move away from the cutting edge, and if quantities justify the cost then a drill jig would be ideal.

Special Considerations

Due to the great range of thermoplastics each having its own pro-perties, it is often necessary to apply special techniques in order to get the best results from machining. For example, it is necessary to anneal rigid P.V.C. before machining. This consists of slow heating to 130° C – 135° C which is held for 30 minutes per 5 mm of material thickness, to be followed by slow cooling in air.

Another example is that of machining P.T.F.E. where precision is of importance. This polymer has a unique combination of properties but has the outstanding disadvantage of having a transition point at 20° C. At this temperature it undergoes a dimensional change of 3%. Therefore it must be machined and measured at temperatures above this. Also smoking must not be carried out by anyone machining P.T.F.E. as swarf can break down, at the temperature of burning tobacco, to give toxic vapours which are harmful if inhaled. Operatives should also wash their hands after work.

11.7 THERMOSETTING PLASTICS

These polymers set solid after being melted to a liquid state by heating; hence the term *thermosetting plastic*. This process of solidifying is known as curing. During curing all the small molecules are chemically linked together to form one giant network molecule. Hence they are distinguished from the linear polymers (thermoplastics) by being called network polymers.

The change from the liquid state to solid is irreversible, and further heating results only in chemical breakdown; not melting. All the common thermosetting plastics are synthetic resins like epoxide, melamine, phenol, polyester for example, which in the hard state are invariably glassy and brittle in nature. Their structure is based upon carbon like the linear polymers. The silicones, however, are another family of materials whose structure is based upon silicon, not carbon. They are usually grouped with the common thermosetting plastics, the properties of some of which are given below in the table shown at Fig 11.8. It will be noticed that fillers are commonly used either to improve or modify the mechanical or chemical properties of the resin, or in order to give cheaper material.

Polymer	Tensile Strength	Compressive Strength	Machining Properties	Chemical Resistance
Epoxy Resin (glass fibre filled)	Outstanding	Excellent	Good	Excellent
Melamine Formaldehyde (asbestos filled)	Good	Good	Fair	Good
Phenol Formaldehyde (bakelite)	Good	Good	Fair	Fair
Polyester (glass fibre filled)	Excellent	Good	Good	Fair
Silicone (asbestos filled)	Outstanding	Good	Fair	Fair

Tensile Strength	Excellent	55 N/mm^2
	Poor	21 N/mm^2
Compressive Strength	Excellent	210 N/mm^2
	Poor	35 N/mm^2

Fig 11.8 Table of Properties of Thermosetting Plastics

Thermosetting plastics are processed to their final shape by moulding. This may be followed in some instances by machining. We will consider each of these processes in turn.

11.8 MOULDING OF THERMOSETTING PLASTICS

Compression Moulding

This is the traditional moulding method which has been used in the plastics industry for many years. The process is superficially similar to the hot pressing of metals in so far that the plastic is placed into a heated mould on a vertical press and subjected to pressure. The main difference is that a thermosetting plastic is moulded at melting temperature whereas a metal is hot pressed at sintering temperature. The plastic is held in the mould from the liquid state until the curing stage is over when polymerization is complete. It should be noticed that compression moulding would be impracticable for thermoplastics which have no curing stage during which a chemical reaction takes place; because the mould would require cooling each cycle in order to allow the component to harden sufficiently for extraction.

The process cycle time obviously depends upon the curing time, and with modern plastics this has been considerably shortened. Vertical hydraulic presses are usually used up to 0·5 MN capacity having an upstroke or downstroke action. Pressure applied to the plastic in the mould is to the order of 15 N/mm², and the curing temperature is to the order of 200° C. Mould heating can be successfully accomplished using steam, gas or electric resistance heating.

Open Flash Mould

Moulds, which may be single or multi-cavity, usually consist of a female and male die giving the component its exterior and interior shape. The first compression moulds were filled with plastic powder by hand, and allowance had to be made for inaccurately measured excess plastic to leak out of the mould in the form of *flash*. Hence, an open flash mould of the type shown in Fig 11.9 must be used in such a case.

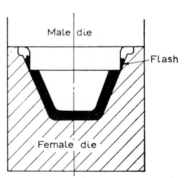

Fig 11.9 Open Flash Mould.

Allowance must be made in the mould size for shrinkage because the

ratio $\dfrac{\text{Volume of powder}}{\text{Volume of moulding}}$ can vary between $\dfrac{3}{1}$ and $\dfrac{10}{1}$.

Closed (Positive) Mould

Compression moulding is now more often carried out using closed moulds which give no troublesome flash upon the component. They require an accurate amount of plastic to be placed in the mould. This can best be done by using a preforming machine which compresses the plastic into pellets of tablets of the appropriate size and weight. One or more pellets are then fed into the mould cavity giving the exact volume of plastic required to fill the mould. A diagram of a closed mould is shown in Fig 11.10.

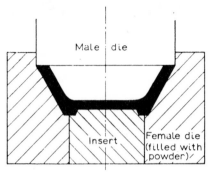

Fig 11.10   Closed (Positive) Mould.

In addition to preforming, preheating is sometimes carried out in gas or electric ovens in order to partially soften the pellets. This is done for two reasons: (i) To reduce the cycle time and increase the production rate by reducing the curing time in the press. (ii) To give easier flow characteristics to the polymer in the mould thus preventing damage to delicate moulds and small pins. This idea has led to the development of transfer moulding which can be regarded as a mixture of compression moulding and injection moulding.

Transfer Moulding

This process is a modification of compression moulding in which the plastic, in the form of pellets, is placed into a separate chamber from which it is transferred by injection into a closed heated mould. Today, high frequency preheating has been perfected by radio frequency methods. This is often used to improve both the surface finish and the cycle time. Fig 11.11 shows a transfer mould. It will be noticed that the configuration lends itself to injection by a central plunger into several moulds located around it.

The big advantage of transfer moulding is that there is little pressure inside the mould cavity until it is completely filled. Only then is full fluid pressure transmitted. This then makes the process ideal for the moulding of delicate and intricate shapes. A simpler arrangement than that shown at Fig 11.11 is to allow the heated liquid polymer to flow under the force of gravity into the mould through suitably placed gates. The plastic is then moving under the laws of fluid flow and there are no forces tending to distort any of the mould features. The charging chamber will then have to be above the mould as shown in Fig 11.12.

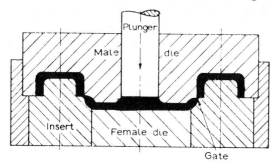

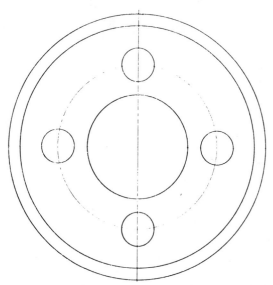

Fig 11.11   Multi-transfer Mould.

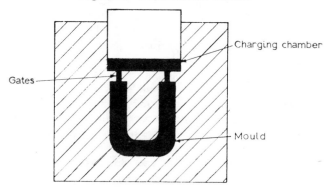

Fig 11.12   Transfer Mould.

Any fillers which are being added to the thermosetting polymer are introduced at the moulding stage.

Injection Moulding of Thermosetting Plastics

The moulding of thermosetting plastics by some form of the injection process has been practised for some years. In principle it is similar to the injection moulding of thermoplastics but does not require a cooled mould in order to harden the polymer.

The operating cycle is as follows:

i) A measured charge of resin is fed into the heating chamber prior to injection. The chamber temperature is to the order of $60°$ C – $100°$ C depending upon the material being processed, and has a water jacket around it to prevent the plastic pre-curing by over-heating whilst it is in the chamber.

ii) The plastic is injected into an electrically heated, locked mould maintained at a temperature of $150°$ C – $180°$ C. The injection pressure is to the order of 70 N/mm^2 – 90 N/mm^2. A reduced pressure is maintained on the resin in the mould during the early stages of the curing process.

iii) The mould is opened and the component ejected after curing is complete.

The mould is closed and re-locked, and the cycle repeats.

Screw injection moulding machines are now successfully being used for the injection moulding of thermosetting plastics. The advantages over a ram type machine are as for thermoplastic injection moulding as outlined in Section 11.3. The machine is generally similar to that shown in Fig 11.3, but the mould is not water cooled. A further major difference is that the screw also acts as a ram in order to maintain pressure on the resin in the mould at the curing stage. To this end the screw is flat faced, and can be moved forward axially like a ram to maintain pressure.

Thermoplastic injection moulding is a high speed production process, but of course thermosetting plastic injection moulding is relatively a much slower process.

11.9 MACHINING OF THERMOSETTING PLASTICS

The machining of thermosetting materials is not widely practised. Comparison of the tables shown in Fig 11.1 and Fig 11.8 reveals that most of the thermoplastics have excellent machining properties whereas the thermosetting plastics do not in general have good machining properties. However, laminated sheets made from a thermosetting resin such as phenol formaldehyde compressed under heat with layers of reinforcing fillers such as paper, cotton, glass or asbestos sheets are often machined to their final shape instead of being moulded. Fillers, which are widely used with thermosetting plastics, greatly influence the behaviour of the material during machining. This adds one more complication to the process and means that the best cutting conditions may have to be found by trial and error. If fillers are used it will be found that tungsten carbide is the only cutting tool material that will retain its cutting edge over an extended production run. Again, a sharp cutting edge with adequate clearance is essential.

The table shown at Fig 11.13 gives a guide to tool angles, speed and feed for one unfilled thermosetting plastic.

| Material | Rake | Clearance | Turning | | Milling | | Drilling up to 25 mm dia. | | Band Sawing |
			Cutting Speed (m/s)	Feed (mm/rev)	Cutting Speed (m/s)	Feed (mm/s)	Cutting Speed (m/s)	Feed (mm/rev)	Speed (m/s)
Phenolic Laminate	0 to 30°	15°	1	0·1 to 0·25	7·5	up to 4	3·5 to 50	0·1 to 0·25	5 to 37·5

Fig 11.13 Tool angles, speeds and feeds for Machining an unfilled Thermosetting Plastic.

P.F. laminates, such as Tufnol, are the plastic materials most comparable with metals. They are supplied in sheet, rod or tube form and generally may be treated as a common brass for the purposes of machining.

Some of the more important machining processes for the P.F. laminated resins will be considered in turn.

Turning

Generally, tools with no top rake and 15° clearance will be found satisfactory, although the rake may be increased up to 30° for heavy cutting. Vee steady box tools are recommended on a capstan lathe. A very fine surface finish can be obtained with fine feeds. No coolant is required, although a small quantity of mineral oil may be found helpful in producing a fine finish.

Milling

P.F. laminates can be milled very easily, particularly by vertical milling at high speeds, or upon woodworking routing machines. High speed steel milling cutters should be used having spiral teeth preferably of fine rather than coarse pitch. Milling is ideal for removing large amounts of material.

Drilling

High speed steel twist drills having the standard point angle of 118° can be used. High spindle speeds will be found best for small hole production. Care must be taken when the drill breaks through as the hole edge easily breaks. Therefore a wooden backing block under the plastic should be used. Accurate reaming will be found possible where the hole size is important.

Use of the *woodpecker* technique is recommended as for thermoplastics in order to clear the swarf which otherwise will quickly clog the drill flutes.

Again, grinding will be found to have no place in the machining of plastics. Also, it will be found that glass fibre filled resins represent a special problem as they are extremely abrasive and rapidly dull cutting tools. High tooling costs make the machining of such materials an uneconomic proposition. Although P.F. laminates are the only thermosetting plastics to be machined to final shape in any quantities at the moment, new resin laminates are being produced which will be just as easily machinable. One such material is an epoxide laminate which is said to machine like P.F. laminate but to have the mechanical properties of a glass fibre filled resin.

11.10 OTHER PROCESSING METHODS FOR PLASTICS

The processes mentioned so far for the production of plastics, viz., moulding, extrusion, casting and sheet forming, have some similarity to a metal forming process. There are other processes however, which have been developed exclusively for the production of plastic articles and only bear superficial resemblance to any other process. We will consider one or two of these processes now.

Calendering

This is an intermediate process in which extruded thermoplastic sections are reduced to sheets for example which may (or may not) then be formed to final shape by vacuum forming. As the plastic is being worked to shape at an intermediate stage, calendering can only be used for thermoplastics and not thermosetting plastics.

The process is similar in one way to the rolling of metal [Fig 11.14(a)] in that the material is compressed between rolls and emerges in the form of sheet. In every other way it is quite different because (i) the thermoplastic material swells appreciably after it has passed through the position of minimum thickness which is directly on the vertical axis of the rolls [Fig 11.14(b)], and (ii) the pre-calendered material is

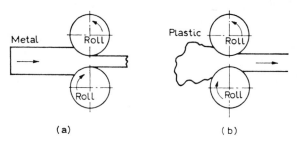

Fig 11.14(a)   Rolling Metal Sheet   **14(b)**   Calendering of a Thermoplastic

not in sheet form but is in an extruded form of any given shape, or even in an unprocessed mass of indefinite shape.

P.V.C., for example, is calendered into the well known transparent film widely used for packaging.

Rotational Moulding

This is a method of moulding thermoplastics which utilizes heat only, unlike injection moulding for example which utilizes both heat and pressure. Relatively, rotational moulding is in its infancy, but is already used extensively for the production of toys such as beach balls, hobby horses, boats, etc., in P.V.C. Polyethylene components such as containers, ducting, etc., are now successfully produced by the process, including large components like laminated petrol tanks for motor cars made from polythene (outer shell) and nylon (inner shell). Developments are taking place to enable a greater range of plastics to be used, and given the right material it is anticipated that a one piece motor car body shell could be produced in plastic by rotational moulding.

The process cycle consists of:

1) A female mould is charged with the exact weight of polymer powder required for moulding.

2) The mould is closed, then rotated simultaneously about two axes perpendicular to each other as heat is applied. This causes the powder to sinter against the mould walls, building up the wall thickness of the component.

3) At the end of the heating and sintering process cooling takes place while the mould is still rotating.

4) Rotation is stopped and the moulding is removed.

Several methods have been developed most of which depend upon using three moulds, one for each stage of the process, i.e., unload and reload mould 1, heat mould 2 and cool mould 3. Cooling is done by applying cold water and air to the outside of the rotating mould. The disadvantage of this system is that the three stages are not equal in time therefore there is idle time at one or two of the stages giving a comparatively slow production rate. At the time of writing it is believed that the latest equipment developed on these principles has none of the disadvantages mentioned. It is differentiated from the earlier process by being called the high speed rotational moulding process.

An unconventional system which seems to hold great promise is the jacketed mould process. This is shown diagrammatically in Fig 11.15.

The female mould is made with a jacket surrounding it. Hot and cold oil are pumped alternately through the jacket to exhaust; the hot oil heating the mould and the cold oil cooling it as it rotates. This is a very compact unit which does away with the need for an oven.

It should be noted in the process of rotational moulding that because only a female mould is used, the component can be held to a definite shape and size on the outside (mould) surface only. The internal surface of the moulding will take up the general shape of the outside surface, the wall thickness being held to a tolerance of $\pm$ 0·25 mm.

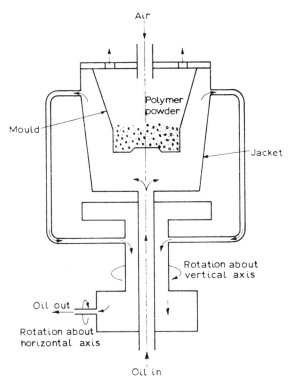

Fig 11.15 Jacketed Rotational Mould.

Expanded Plastics Moulding

Thermoplastics may be produced in cellular form. Examples of this are expanded polystyrene packaging, the well known thermal insulation (in sheet or tile form), or buoyancy objects such as life rafts. The density of these objects is to the order of 25 to 35 kg/m³. There is more than one way of foaming the material in order to expand it before moulding it to a final shape. Let us consider the common method applied to expanded polystyrene products.

The process can be divided into three stages:

1) The raw material consisting of plastic beads between 1 mm and

0·5 mm in size, are expanded to many times their original size under the influence of steam heating. A hydrocarbon contained in the polymer is vaporized into small bubbles which cause the expansion of the plastic. This stage may be done in batches or continuously as shown in Fig 11.16.

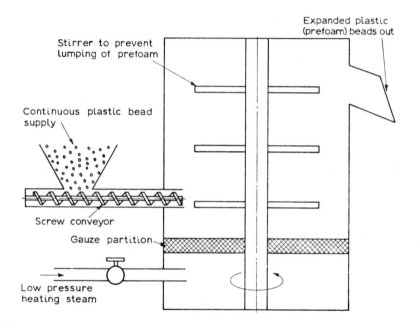

Fig 11.16   Continuous Plastic Prefoamer.

2) The expanded plastic beads (called prefoam) are dried.
3) The prefoam is injected by air into the mould cavity and is fused by heat to its final shape. Before removal it must be set, and for this purpose the moulds are cooled by water. The component is ejected after the moulds open, and is finally dried before use.

The moulding is usually carried out on a hydraulic press of vertical or horizontal configuration. Modern machines may be fully automatic. A diagram of the set-up is shown at Fig 11.17.

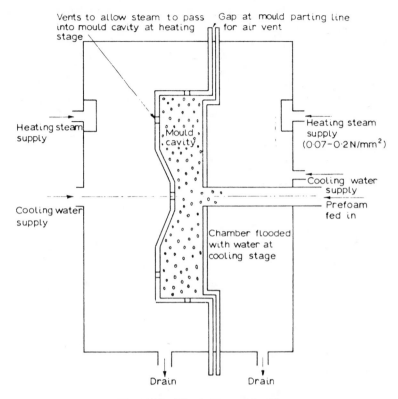

Vents to allow steam to pass into mould cavity at heating stage

Gap at mould parting line for air vent

Heating steam supply

Heating steam supply (0·07–0·2 N/mm²)

Mould cavity

Cooling water supply

Cooling water supply

Prefoam fed in

Chamber flooded with water at cooling stage

Drain   Drain

Fig 11.17 Plastic Foam Mould.

11.11 PLASTIC COMPONENT DESIGN

On occasions it is possible to replace simple metal parts by an equivalent plastic part to some advantage. The fact must be faced however, that when designing with plastic rather than metal materials, quite different design considerations apply. Plastics should not be treated like metals because their properties are quite different. Plastics show pronounced creep at room temperature, they suffer from stress-relaxation, fatigue, and thermal distortion, and have a very narrow working temperature range. The maximum working temperature for most thermoplastics is to the order of 100° C, and for thermosetting plastics is to the order of 250° C. A further complication is that the behaviour of plastics is non-linear. To sum up there is a complex time-temperature-stress-strain relationship which affects the mechanical behaviour of all plastics, and much more data is required by the designer than at present exists.

On the other hand, if the production engineer treats the field of plastics as a separate subject worthy of detailed study both in design and production he will find many advantages. These will include fast and cheap production, durable and attractive products in varied colours and finishes, high strength/weight ratio, good electrical insulation properties, corrosion and chemical resistance, and built-in bearing properties. One big limitation which exists at present is that of size. Much research is needed before a large structure like a motor car body can be made in an acceptable plastic, more efficiently and cheaply than a metal body.

In this section we can now briefly consider some of the factors to be taken into account when designing plastic components.

a) Choice of Material
The first requirement is that the material chosen (as in every field of design) meets its functional specification and is capable of withstanding the working conditions imposed upon it during service. For example, a small cam may be required in large quantities as a moulding, to be used at room temperature with little frictional resistance, and to be assembled with self tapping screws to another part. Nylon, polycarbonate or acetate would be suitable, or any of these polymers combined with a filler such as glass fibre. One of these materials probably has the best all round combination of properties to suit the specification, and the final choice can only be made as a result of one's knowledge and experience of the plastic through testing and usage. The final choice will then depend upon the cost, and if the quantities warrant it an analysis may be carried out.

b) Designing to Facilitate Production
It has already been stressed that the important point is to design the part bearing in mind the different properties of plastic compared to metal. In addition, the plastic part can often be moulded to its final shape in one piece (if quantities permit) with no further processing or machining required. The two following examples illustrate these points.

Example 11.1
The component shown at Fig 11.18 is fabricated from mild steel plate gears, turned spindle and screws. It would cost more to machine it from solid with consequent high metal wastage.

The equivalent component designed as a one piece moulding in nylon is shown at Fig 11.19.

Note that two bearing diameters, A and B, are now included in the spindle, B being a greater diameter than A. This is because the production of accurate, undistorted holes in plastic articles is difficult. Therefore, the bearing arrangement shown allows for a much greater degree of hole distortion without destroying the serviceability of the component. This illustrates a further important issue, viz., that it seldom pays to treat each plastic component in isolation but the total design effect should always be considered. The reader may note that all-plastic food mixers, clocks, gauges for example are now available each of which functions equally as well as its metal counterpart. Indeed, plastic moving parts run with a much greater degree of quietness.

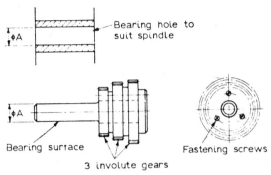

Fig 11.18   Mild Steel Multi-gear.

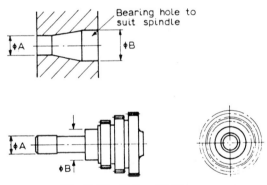

Fig 11.19   Nylon Multi-gear.

Example 11.2
The component shown at Fig 11.20 is a machined, cast iron two part casting held together with bolts.

The equivalent component designed as a two part moulding in poly-carbonate is shown at Fig 11.21. The parts are fastened together with self tapping screws.

Note that in the interests of moulding technique and of stiffening the plastic container in order to make it more rigid several design changes have been made. These are (a) the parting line has been brought to the centre line, (b) large corner radii are used, (c) a constant thickness section is provided, (d) ribs are used to support the long bosses.

Adequate taper (draft) would also be provided where possible, and a male register (spigot) would be provided on the part 2 in order to easily locate part 1. Very precise location can be achieved by a tongue and groove arrangement.

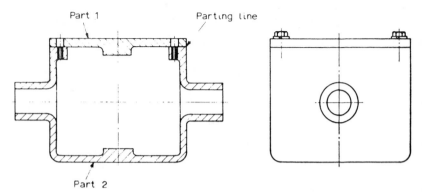

Fig 11.20   Cast Iron two part Component.

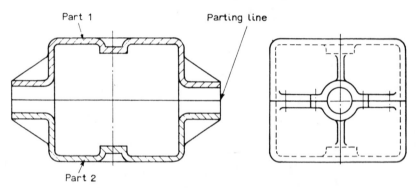

Fig 11.21   Polycarbonate two part Component.

c) Designing to Facilitate Assembly

Plastic components can be easily assembled by a variety of methods, some making use of their unique properties. A few illustrations of assembly design techniques for plastic parts are given in Fig. 11.22.

a) Tapped metal insert located in mould to become an integral keyed part of plastic component.

b) Counterbore above cored hole to maintain screw in upright position before it is driven home. It also stops possible cracking of boss as screw is started.

c) Ultrasonic tools can be used to enter screws or inserts into certain plastic resins without rotation. The plastic flows around the screw without the plastic being cut. The screw can subsequently be screwed out leaving a full form thread.

d) Plastic parts can be rotated together under light load until sufficient heat is generated to cause the two parts to weld together as one.

e) A heated punch can be used to swage over the shank of a plastic part giving a perfect riveted head.

The majority of plastics can be joined by adhesives. Some, such as acrylic, can be instantaneously joined by using an appropriate solvent.

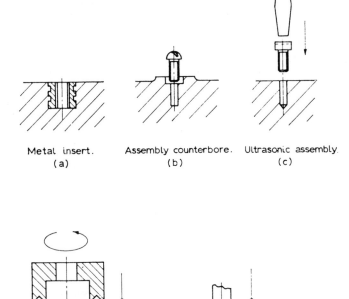

Metal insert.
(a)

Assembly counterbore.
(b)

Ultrasonic assembly.
(c)

Friction welding assembly.
(d)

Hot swaging assembly.
(e)

Fig 11.22 Plastic Assembly Techniques.

11.12 MOULD DESIGN

The greatest number of plastic moulds are made for injection moulding, and in principle the problems are very similar to those encountered in pressure die casting of metals with one or two differences. The low viscosity of molten aluminium means that moulds can be filled under pressure with this metal in a fraction of a second. The high viscosity of melted thermoplastic on the other hand means that mould filling times are very much slower, and pressures have to be much greater. Also metal processing moulds must be water cooled to extract heat from the mould, whereas most plastic processing moulds need not be cooled.

Like all moulding processes, the quality and accuracy of the finished component depends largely upon the quality and accuracy of the mould cavity. A highly polished surface and adequate draft are necessary to give a good finish, and to assist ejection of the component. Moulds can be manufactured from cast iron or mild steel for short runs, case hardened nickel chrome steel for medium production runs, and non-shrinking

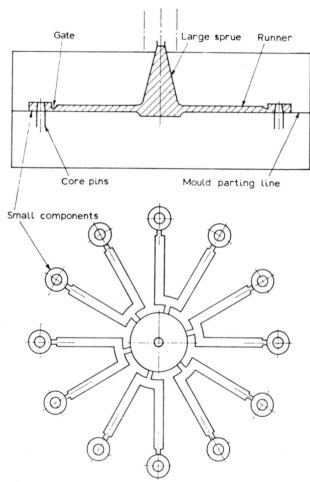

Fig 11.23(a) Multi-impression Mould having large Sprue and Runners.

alloy die steel for extended runs. Cavities can be shaped by conventional machining techniques such as die sinking or spark erosion; or by hobbing. (The last mentioned consists of forcing a hardened, male facsimile of the cavity called a hob into the soft piece of die steel. This is done under considerable pressure, the steel die block afterwards being stress relieved and case hardened.)

Single cavity moulds are used for greatest accuracy, but multi-cavity moulds are used for high production. As shown in Fig 11.23(a) the sprue and runners may be much greater in weight than the components. Again, depending upon circumstances, the opposite may be true as shown in Fig 11.23(b).

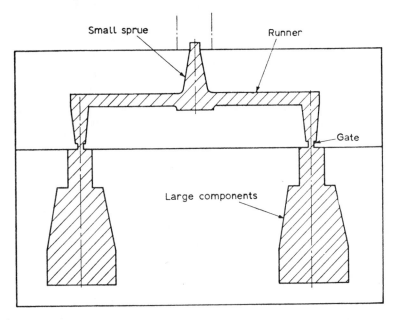

Fig 11.23(b) Double impression Mould having small Sprue and Runners.

Finally, one important point should be mentioned about the design of plastic processing moulds compared to others. Shrinkage can often be large and unpredictable, and obviously must be allowed for in the interests of accuracy. Tolerances of $\pm$ 0·025 mm may be difficult to achieve. Cavities can occur in the component due to excessive shrinkage, and one way of overcoming this is to use push back core pins. This technique is illustrated in Fig 11.24.

The pin is shown in its normal position. When a shot of plastic enters the cavity the core pin will be pushed back under pressure; as the plastic cools and shrinks the pin re-enters the mould cavity, pushing into the component a controlled amount hence preventing sinking and cavitation and a giving a cored hole. As the important face of the component in this case is at the front, a flat face is assured.

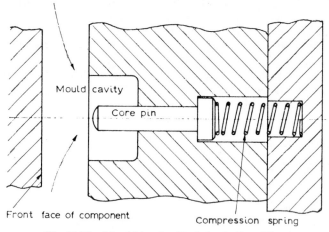

Fig 11.24   Mould having Push back Core Pin.

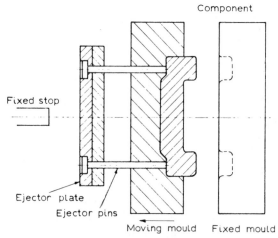

Fig 11.25   Mould Ejection Mechanism.

Components are ejected from the mould as it opens using the orthodox method of round pins bearing upon unimportant areas of the component faces. This is shown in Fig 11.25.

As moving mould opens the ejector plate hits the fixed stop. The mould continues to open and the pins eject the component from the mould. Alternatively the ejection may be controlled by hydraulic or pneumatic means.

Exercises 11

1. (a) What is a thermoplastic?
(b) What is a thermosetting plastic?

2. With the aid of sketches compare the injection moulding of a thermoplastic with the pressure die casting of aluminium using the cold chamber process. Contrast working temperatures and pressures.

3. Compare the extrusion of a thermoplastic with the hot extrusion of aluminium. What are the differences in working pressures and temperatures?

4. Corrugated sheets of P.V.C. are required in large quantities. Briefly outline the processes required for their production. Use simple line diagrams as a means of illustration.

5. A plain P.T.F.E. bush, 35 mm O/dia. × 15 mm bore dia. × 40 mm long is to be produced upon a capstan lathe in small batches. Show a diagram of the set-up and give detailed sketches of the tools used indicating tool angles and tool material. Estimate the standard time for the operation. The nearest stock size of P.T.F.E. rod available is 45 mm dia.

6. Compare and contrast the following methods of moulding thermoset resins: (i) compression moulding, (ii) transfer moulding, (iii) injection moulding.

7. A simple gear pump consists of a housing, two involute gears and spindles, the housing being split to facilitate assembly. In service, one spindle is coupled to an electric motor spindle. Produce design details for the pump to be manufactured (a) from all metal parts, and (b) from all plastic parts. In each case indicate the material to be used for each part.

8. A thermoplastic component is to be injection moulded (a) very accurately in small batches, (b) in very large quantities with accuracy of secondary importance. What would be the main differences in the mould design for application (a) and (b) respectively.

9. Compare two common thermoplastics with a plain, low carbon steel with respect to: (i) mechanical properties, (ii) electrical properties, (iii) corrosion resistant properties, (iv) temperature range, (v) range of finishes available, (vi) economics of processing and machining.

Which of these materials would you choose for a rocker box cover on the cylinder head of an overhead valve, internal combustion engine for a popular saloon car?

10. With respect to the economics of production of motor car bodies it has been said: 'When providing the dies for a steel body the capital cost is the same whether you are going to make 20 a week or 2000 a week. With plastics it is different . . . '

Discuss the cost situation and state where you think the fundamental difference lies, if any, between producing plastic or metal car bodies. You should consider both small quantity and large quantity production.

Further Reading

1) Alexander J. M. and Brewer R. C. *Manufacturing Properties of Materials.* D. Van Nostrand Co. Ltd.

2) Saunders D. W. and Stuart J. M. 'Plastics for the Engineer.' *Journal of the Institute Mechanical Engineers*, 1967.

3) Devereux A. R. 'A Comparison of the Properties of Conventional and Plastics Materials.' *Journal of the Institute of Production Engineers*, 1967.

4) Sharp H. J. 'Manufacturing Properties of Plastics Materials.' *Journal of the Institute of Production Engineers*, 1967.

5) Ryall H. 'Precision Mechanisms via Injection Moulding.' *Journal of the Institute of Production Engineers*, 1967.

6) Turner L. W. and Johnson L. W. 'A Review of Recent Developments in Injection Moulding and Blowing Equipment.' *Journal of the Institute of Production Engineers*, 1967.

7) Burges-Short M. G. 'Fast Cycle Thermoset Moulding by Screw Injection.' *Journal of the Institute of Production Engineers*, 1967.

8) Macknight-Thomson I. D. 'The Development and Application of the High Speed Plastics Rotational Moulding Technique.' *Journal of Institute of Production Engineers*, 1967.

9) Kent J. M. 'Machinery for the Production of Expanded Polystyrene Package Mouldings.' *Journal of Institute of Production Engineers*, 1967.

10) Fagence S. W. 'Thermo-forming Materials, Machines and Methods. *Journal of Institute of Production Engineers*, Part I May 1968, Part II June 1970.

11) Holmes-Walker W. A. and Smith I. B. 'The Competitiveness of Plastics.' *Journal of the Institute of Production Engineers*, May 1975.

12) 'Working with Plastics.' *Journal of the Institute of Production Engineers*, May 1979.

CHAPTER 12

Press Tools

12.1 FACTORS AFFECTING PRESS TOOL DESIGN

IN THE manufacturing section of the engineering industry, metal articles can be worked to shape either by metal cutting, or by metal forming. Metal cutting can often be wasteful because, on the average, 40% of the original component material is removed by expensive machining operations to become scrap. This scrap material will then on the average be worth 5% of its original value as raw material. In many cases machining operations can be more economically carried out by metal forming. There has been a move in recent years to greatly extend the scope of metal forming, which includes such processes as hot and cold extrusion, hot and cold forging, and press tool work.

It is interesting to carry out a break-even cost analysis (see Section 1.3) upon a manufacturing operation where metal cutting or metal forming are possible alternative processes. Often the forming process requires expensive dies and fixed costs are higher; material wastage is negligible, labour costs are low and hence variable costs are lower. The metal cutting process is often lower on fixed costs, but higher on variable costs. Hence, the forming process will be more economical when quantities required are large.

In this chapter we shall be concerned solely with the use of presses and press tools as a means of cold working metal objects into shape. (We shall not cover press brakes or turret presses which may be regarded as more specialized topics.) Most of this type of work is carried out upon ductile metal in sheet or strip form of relatively thin section. It represents an important part of the manufacturing industry, being used for the cheap production of large quantities of components, such as motor car bodies, electric motor parts, domestic electrical articles, etc. Orthodox presses (usually vertical) are used, and may be hydraulic or pneumatic powered, or may be mechanical crank presses. For very light work, hand presses such as fly presses, are satisfactory. The metal if in coiled strip form, may be fed automatically into the press tool by power rolls or power slide, or may be hand fed by the operator. If the metal is in some other form, such as a sheet or partially formed shape, it may be located in the tool by mechanical hands which have gripping fingers, or locating pans which drop the metal part into the correct position. Again, an operator(s) may hand feed the part. The mechanical feed or location devices must of course be synchronised to operate every time the press ram lifts the top tool clear of the bottom tool. Where the press is hand fed, stringent safety precautions must be taken to ensure that the operators' hands cannot be trapped in the press tool. Efficient guards must be provided which are completely foolproof, and it should always be remembered that a press is potentially a very dangerous machine.

In order to appreciate press tool design, it may help to briefly reconsider the elementary principles of metal plasticity. Standard tensile or compression tests which cold work the specimen being used are an ideal means of obtaining data about the plastic range of metals. Consider Fig 12.1 which shows the results of a tensile test upon a relatively ductile material, such as a low carbon steel.

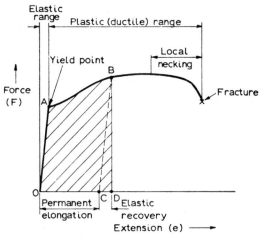

Fig 12.1 Tensile Test of a Ductile Metal.

The metal is elastic up to point A and will return to its original size if the force is withdrawn. If the force however, is increased to point B before being withdrawn, the force-extension graph follows line BC, parallel to line AO, as the force is removed. The test piece will then be permanently extended by amount OC, and will not return to its former size. CD represents the elastic contraction (recovery) which occurs as the force is removed.

Area OABD represents the work required to cause deformation OC.

If the overstrained material is again subjected to a tensile force upon a testing machine, we shall plot an entirely different force-extension graph than we first derived. This second graph will now have its origin at C (instead of O), its yield point approx. at B (instead of A), and its breaking point at approx. the same point as would have occurred if the first test had been completed to failure. In effect, the original piece of metal in being cold worked to point B well above the yield point acquires a new set of properties. These new properties result in a different force-extension graph being derived if the metal is re-worked. This is the most important first effect of cold working, which means that a cheaper material can be specified for a cold forming operation, the component finishing with new superior properties comparable to a more costly material.

The results of cold working a metal to point B well within the plastic range of the metal can be summarised as follows:

a) The yield point, and hence the stress at yield point is raised, where

$$\text{stress at the yield point} = \sigma_y = \frac{\text{Force at yield point}}{\text{Cross sectional area}}$$

b) The ductility is lowered, and hence the elongation % is reduced,

$$\text{where elongation} \% = \frac{\text{extension}}{\text{original length}} \times 100$$

Should high ductility be the important property, such as in a deep drawing operation which must be carried out in several stages (see Fig 5.9), then the metal must be annealed after cold working to restore it to its original state. From Fig 12.1 it can be seen that if no annealing takes place, then between points A and B, each successive increment of elongation will require an increasing increment of work. In other words, as cold working proceeds, the resistance to deformation rises steadily.

Also reference to Fig 12.1 shows another important factor relating to press tool work. In cold working there will be minor changes in dimensions of the workpiece when the work is removed from the tool. This is the elastic recovery of the material shown in the graph due to the release of stresses causing the deformation. The work after removal from the press tool will spring back from the die shape to take up a different shape. This must be allowed for in the tool design.

Figure 12.2 shows an important relationship between the grain size of the metal after annealing and the amount of cold work applied to the metal before annealing.

This graph shows that there is a critical amount of cold work at which point the grain size of the metal will be coarse. This critical value is shown at point X in Fig 12.2 and is about 10% for a low carbon sheet steel. If a workpiece made from this material receives approx. 10% deformation during cold working, and is then annealed in order to restore ductility say, then the coarse grain will result. This spoils the finish of the article because the large grained structure shows up in an 'orange peel' effect on the surface of the work. Where heat treatment is to follow cold working, the critical amount of cold working should be avoided.

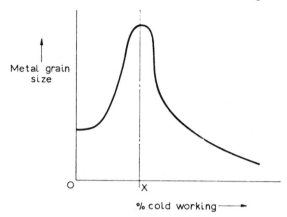

Fig 12.2

One other factor to consider in press tool work is the direction of rolling of the strip to be worked in the press tool. Cold rolled sheet or strip is used which has been rolled in a certain direction thus giving the strip directional properties. It will be found that bending can be carried out more successfully *across the grain* (direction of rolling), than *along the grain*. Components which are produced from tools which effect bending along the direction of rolling will almost certainly crack during the bending operation.

There are three different ways of cold working sheet metal in press tools. These are:

1) *Shearing*. In this case the required shape of work is sheared from the metal strip, the metal being deformed to shear failure. There are three variations of shearing, viz.,

a) Blanking, in which a blank is punched from the strip, the blank removed by the punch being the required article. The metal left is waste. The die is made to the required shape and size, i.e., the punch is made smaller than the die by the amount of clearance required.

b) Piercing, in which the blank punched or pierced from the metal strip is waste, the hole left in the strip being required. The piercing punch is made to the required shape and size, the clearance being added to the die.

c) Cropping, in which the piece blanked from the strip is waste, the strip left with the cropped ends being the required part. A cropping tool is, in principle, a blanking tool.

Each of these operations is shown at Fig 12.3(a), (b) and (c) respectively.

In Fig 12.3(b) it can be seen that the operation is one of piercing three holes followed by the strip moving forward one pitch. Then a blanking operation follows to give a blanked and pierced component. Piercing is most commonly carried out in conjunction with blanking.

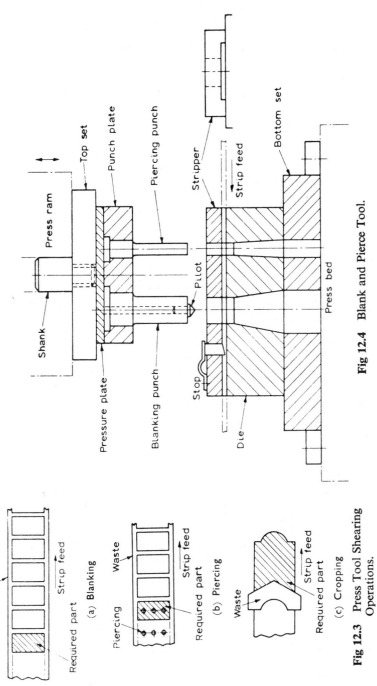

Fig 12.4 Blank and Pierce Tool.

Fig 12.3 Press Tool Shearing Operations.

(a) Blanking

(b) Piercing

(c) Cropping

Figure 12.3(c) shows a cropping operation which is used when the component is relatively long, and the width is sufficiently accurate without blanking. This is economically good sense where it can be done, as the tool is cheaper than a full blanking tool.

2) *Bending.* This is carried out on blanks, strip, sheet, rod or wire and consists of local deformation, as opposed to a change of shape of the complete article. Forces must be high enough to cold work the material within the plastic range.

3) *Drawing.* This is carried out on blanks, and involves considerable deformation or a complete change of shape of part. Again the deformation must be carried out in the plastic range. As stated earlier, deep drawing may require several drawing stages with interstage annealing.

In the next section we will consider each of these methods of press work, and the principles of design of a typical tool of each type.

12.2 SHEARING

Figure 12.4 shows a blanking and piercing press tool. This is a simple shearing tool, and the punches and die can be of any required profile.

The tool shown has the main features of any press tool which we will examine in more detail.

Punch

Made from a non-shrinking, non-distorting alloy tool steel, such as a high carbon, high chromium alloy steel. Punches are hardened and ground, and are held in a punch plate. The shanks may be a drive fit in the punch plate as shown. Alternatively, where there are many piercing punches say, which require precise location in the punch plate to match the die, the punches may be held in place in the punch plate by Cerromatrix or some other low melting point alloy. A little thought will show that this method greatly eases the problems of manufacture of the press tool.

Die

Made from the same metal as the punch, and like the punch is re-sharpened by grinding the top face. Complex die shapes may necessitate the die being made in more than one piece. The die profile is 'backed off' with taper as shown to allow the blanks and piercing slugs to easily fall clear into tote boxes positioned under the press bed.

Stripper

This may be made out of mild steel and is a clearance fit for the punches at the top, and the strip at the side which it guides into the correct position under the punches. The stripper prevents the metal strip lifting up with the punch as the ram returns. It also provides a means of housing the stops.

Stop

There are many forms of stops, the one shown being a simple spring loaded type. As the strip is pushed under the stop it rides up over the strip to drop into the space left by the blank. If the strip is then pulled back against the stop (against the direction of feed), the strip is correctly located for the next shearing operation, leaving the minimum of waste metal between the blanked holes. This feeding operation up to a stop can be carried out at high speed by the operator, and a well designed stop must be simple and efficient in use.

More complex tools may require more than one stop, and arrangements can be made to operate stops by the movement of the press ram if desired. Automatic feed devices do not require stops as the strip is automatically moved along the correct pitch length, between each stroke of the press.

Pilot

These are held in the blanking punch and are a clearance fit for the pierced hole. As the ram descends, the pilot locates the previously pierced hole and positions the strip more precisely under the blanking punch. Pilots are used where the relationship of pierced holes to a blank profile must be precise; therefore a stop initially positions the strip, but the final positioning depends upon the pilot. It can be seen then that this technique is ideal for a roll feed press which has no stops upon the tool.

Sets

These may be made from mild steel as they are bolsters to which the die and punch assemblies are screwed. The top set holds the shank by means of which the punch assembly is located and held in the press ram. The bottom set is bolted or clamped to the press bed in the correct relationship to the top set so that the axes of the punches and die holes are in line.

Cast iron die sets are commercially available which are made complete with guide pins and bushes. Hence the top set can always be located in exactly the same position relative to the bottom set. The press tool maker then ensures that the punch and die are always a perfect fit when the tool is closed. This makes the setting operation on the press quick and easy, because the complete tool is simply mounted and fastened on the press, no locating between punch and die being necessary.

Pressure Plate

This is a hardened and ground steel plate which is inserted between punch and top set, in order to take the impact on the head of the punch as it shears through the strip.

Let us now consider some other important features of sheet metal shearing which affect the design of the tool.

Clearance between Punch and Die

When the punch hits the work metal strip, it penetrates a certain percentage of the strip thickness before the metal shears and the whole blank ruptures. The depth of penetration depends upon the hardness of the work material, being greater for a soft material, and can be seen as a polished rim around the edge of the blank.

The amount of clearance allowed between punch and die also affects the appearance of the blanked edge and the accuracy of the finished blank. The clearance varies with the thickness and hardness of the metal being sheared and will be some value up to 10% of the strip thickness. The edge surface appearance and quality produced upon a standard tool will not be as good as a machined surface, but can be adequate if the correct clearance is chosen. Figure 12.5 shows the effect on the blank edge of allowing insufficient clearance. Note that the clearance is the gap between the adjacent walls of punch and due, i.e., radial clearance for circular punches.

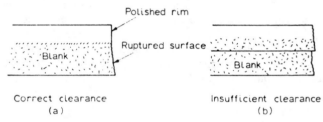

Fig 12.5   Effect of insufficient Clearance.

Force and Work Done required for Shearing

It is necessary to know the force for a particular shearing operation in order to choose a press of adequate capacity for the press tool. The shearing force varies during blanking (piercing or cropping) because of the nature of the process as just described. This is illustrated in Fig 12.6 which shows a force-penetration graph for a blanking operation in which the correct amount of punch and die clearance was allowed.

It can be seen from the graph curve that the force reaches a maximum (F_{max}) as the punch penetrates the metal, then falls away rapidly as the metal ruptures. The graph shown is for a fairly hard steel where $c = 15\%$ say. For soft steel, c might equal 40%, but F_{max} would be much lower. This maximum punch force depends upon the edge area to be sheared and the shear strength of the metal, therefore:

$$F_{max} = \text{(ultimate shear stress of material)} \times \text{(shearing area)}$$
$$= \tau \times \text{material thickness } (t) \times \text{work profile perimeter.}$$
$$= \tau \times t \times x$$

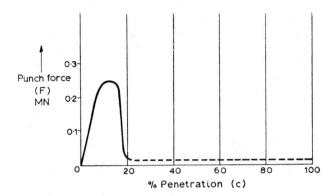

Fig 12.6

The area under the force-penetration curve is equal to the work done during the shearing operation, therefore:

Work done = (maximum punch force) × (% penetration) ×
(material thickness)

$$= F_{max} \times c \times t.$$

This is an estimate of course, and depends upon the assumption that maximum punch force is sustained during complete punch penetration of the work metal.

Example 12.1
Calculate the maximum punch force necessary to blank a steel washer 44·45 mm O/D × 22·23 mm I/D × 1·59 mm thick, if $\tau = 432$ N/mm². Estimate the work done if % penetration is 25%.

Solution

$$\text{Work profile perimeter } x = \pi(44{\cdot}45 + 22{\cdot}23)$$
$$= 209{\cdot}5 \text{ mm.}$$
$$F_{max} = \tau \,.\, t \,.\, x$$
$$= 0{\cdot}432 \times 1{\cdot}59 \times 209{\cdot}5$$
$$= 0{\cdot}144 \text{ MN}$$
$$\text{Work done} = F_{max} \,.\, c \,.\, t$$
$$= 144 \times 0{\cdot}25 \times 1{\cdot}59 = 57 \text{ J.}$$

Shear
It can be seen from the above expression for work done, that if the work is spread over a greater stroke movement of the punch, then F_{max} will be reduced. Hence the tool could be used upon a smaller capacity press (assuming the press shut height and stroke are satisfactory) which

could be both convenient and more economical. This effect can be achieved by grinding either single or double shear upon either the punch or die. Shear is usually applied when thick metal is being blanked, or if the work has an extensive contour. Examples are shown at Fig 12.7.

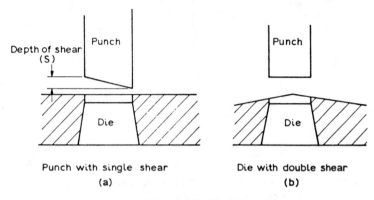

Punch with single shear   Die with double shear
(a)   (b)

Fig 12.7 Shear applied to a Press Tool.

Shear is applied to the punch [Fig 12.7(a)] when piercing or cropping, because the slug which is punched out will be deformed.

Shear is applied to the die [Fig 12.7(b)] when blanking, because the flat face of the punch produces a flat blank without distortion.

The amount of shear (s) to be ground upon the tool depends upon the reduction in punch force required. By a consideration of the amounts of work done with, or without shear, the following expression can be deduced:

$$s = \frac{(F_{max} - F) \times c \times t}{F}$$

This is true for single or double shear.

Example 12.2

In Example 12.1 calculate the amount of shear which must be ground upon the tool, if the maximum punch force is to be reduced to 0·06 MN.

Solution

$$s = \frac{(F_{max} - F) \cdot c \cdot t}{F}$$

$$= \frac{(144 - 60) \times 0·25 \times 1·59}{60} = \frac{33·2}{60}$$

$$= 0·553 \text{ mm}$$

Finish Blanking

The edge finish and accuracy of blanks can be improved by using a technique known as finish blanking. (See reference No. 7 in Further

Reading at end of this chapter.) This technique has been developed by the Production Engineers Research Association (PERA), and its use in industry is developing rapidly. A finish blanking tool is different from a conventional blanking tool in two respects:

1) A very small clearance is used giving a close fitting punch.
2) The top edge of the die is given a small radius.

Smooth edged blanks are produced using this method, but the work done in producing the blank is considerably increased.

12.3 BENDING

Figure 12.8 shows a bending tool for producing a simple component.

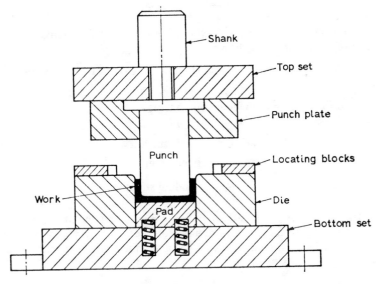

Fig 12.8 Bending Tool.

The metal blank ready for bending is positioned on the die in the locating blocks. The punch descends, and the work is held between the punch face and the spring loaded pressure pad. The die and punch corner radii should be as large as possible to assist forming, and preferably not less than twice the thickness of metal. As with shearing, the punch force varies as the operation takes place. The pressure pad ejects the component after forming.

Force Required for Bending
Allowing for friction, a satisfactory estimate can be obtained of the maximum punch force necessary for bending, by assuming that the bending stress is half the shear stress for the material.

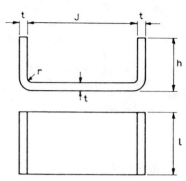

Fig 12.9   Component.

Referring to Fig 12.9 the maximum punch force for completing one bend of the component shown is:

Maximum punch force = (Bending area) × (Bending stress)

$$\therefore F_{max} = l \times t \times \sigma_B$$

For bending the whole component,

$$F_{max} = 2l \,.\, t \,.\, \sigma_B$$

Where t = metal thickness

l = bend length

$$\sigma_B = \text{bending stress} = \frac{\tau}{2}$$

Bending forces are less than shearing loads, and the above approximation gives a generous allowance. However, a more precise value of punch forces could be obtained by a detailed consideration of the bending moments involved.

Planishing
An important consideration which often arises when bending is that of planishing the component. If this is required, the wall clearance between punch and die must be made less than t, and the bottom position of the press ram stroke must be set such that the bottom clearance in the die is less than t. Then the component will also be ironed out, or planished, as well as being formed to shape, and the material must be stressed above yield point to achieve this.

The additional punch force necessary for planishing is given by:

Planishing force = (planishing area) × (yield stress)

$$\therefore F_P = l \,.\, J \,.\, \sigma_y$$

The effect of planishing is to smooth and flatten the work and is done to improve the finish and set the bends.

Allowance for Bending

When bending takes place, the outside layers of the work are under tension and are thus lengthened, and the inside layers are under compression and are thus shortened. As the volume of metal remains the same, then the width on the compressed side of the bend increases, while that on the stretched side of the bend decreases. This gives the characteristic deformed shape on bend sides that one gets if bending a flat plate in a vice. This deformation leads to displacement of the neutral axis towards the compressed side, because the neutral axis coincides with the centre of gravity of the work.

To calculate the length of blank required for bending, we must first calculate the length of the neutral axis of the finish formed component. In practice an allowance is then added for bending to compensate for the shift in the neutral axis. These allowances will be found tabulated in tool reference and design books.

Example 12.3

The component shown in Fig 12.9 has the following dimensions: $J = 76 \cdot 2$ mm, $l = 38 \cdot 1$ mm, $t = 3 \cdot 18$ mm, $r = 6 \cdot 35$ mm and $h = 31 \cdot 75$ mm. Calculate (a) the maximum punch force for bending the complete component, (b) the force if planishing is carried out, and (c) the strip length necessary to make the part (ignoring bend allowances). Take τ as 386 N/mm² and σ_y as 278 N/mm².

Solution

a) $F_{max} = 2l \cdot t \cdot \sigma_B$ where $\sigma_B = \dfrac{\tau}{2} = 193$ N/mm²

$\quad\quad = 2 \times 38 \cdot 1 \times 3 \cdot 18 \times 0 \cdot 193$
$\quad\quad = 0 \cdot 047$ MN

b) $F_P = l \cdot J \cdot \sigma_y$ where $\sigma_y = 278$ N/mm²
$\quad\quad = 38 \cdot 1 \times 76 \cdot 2 \times 0 \cdot 278$
$\quad\quad = 0 \cdot 806$ MN

c) Length of vertical sides.
$\quad\quad = 2(h - t - r)$
$\quad\quad = 2(31 \cdot 75 - 3 \cdot 18 - 6 \cdot 35) = 2 \times 22 \cdot 22 = 44 \cdot 44$ mm

Length along bottom
$\quad\quad = J - 2r$
$\quad\quad = 76 \cdot 2 - (2 \times 6 \cdot 35) = 63 \cdot 5$ mm

Length of bends
$$= \pi\left(r + \frac{t}{2}\right)$$
$\quad\quad = 3 \cdot 142 (6 \cdot 35 + 1 \cdot 59) = 24 \cdot 95$

$\therefore$ Total developed length $= 44 \cdot 44 + 63 \cdot 5 + 24 \cdot 95$
$\quad\quad\quad\quad\quad\quad\quad\quad\quad\quad = 132 \cdot 89$ mm

In this example, right angled bends are required. If the bend is formed in a right angled die, then elastic recovery, or spring back will occur after bending, which can cause difficulty in ejection. Also the corner angle will be greater than 90°. Different techniques are used to overcome this such as planishing and setting the bend, or overbending for example.

Other standard forms of bending tools are shown at Fig 12.10, the method used at (b) being similar to a press brake technique.

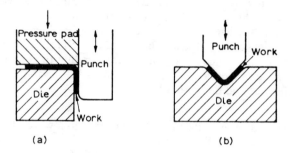

Fig 12.10   Press Bending Tools.

12.4 DRAWING

Figure 12.11 shows a typical drawing tool in which a blank is being drawn into a cylindrical cup. The work is shown in a partially drawn state, the punch not yet having 'bottomed'.

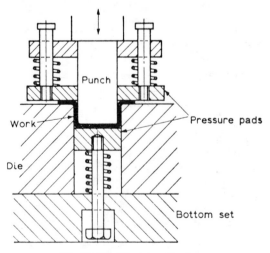

Fig 12.11   Drawing Tool.

The spring (or rubber) loaded pressure pad on the punch keeps the metal tight against the die face, and the blank is *ironed out* as it is drawn over the radiused edge of the die hole. This prevents wrinkling of the cup and keeps the finished edge of the rim straight. The pressure which is applied by this pad is important; no pressure gives heavy wrinkling, and excessive pressure results in the bottom being pressed out of the cup. The optimum pad pressure giving best results varies with the type of work, but will be to the order of 30 to 40% of the drawing pressure. Figure 12.12 shows the stresses which occur as drawing takes place.

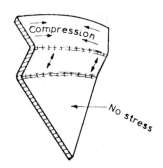

Fig 12.12   Drawing Stresses.

As the blank is crowded into the die with a consequent reduction in diameter, the compressive stress in the undrawn flange will cause wrinkling if this stress exceeds the tensile stress. The correct use of a pressure pad prevents this. If, on the other hand, the force on the blank is such that the tensile stress in the wall exceeds the ultimate tensile stress (σ_u) of the material, then the walls will crack and rupture. However, referring to Fig 12.1, the force must be of such a magnitude that the component is worked within the plastic range of the material. This leads to the conclusion that, for deep drawing particularly, the material must be very ductile having a low ratio: $\dfrac{\sigma_y}{\sigma_u}$

The die pressure pad shown in Fig 12.11 supports the work and also acts as an ejector.

The restriction mentioned above on tensile forces means that there is a limit to the amount of drawing which can be attempted in one operation. It has been found that the drawing ratio $\dfrac{D}{d}$ should not exceed a value of 2 (where D = blank diameter and d = cup diameter) for most materials, although this depends upon the value of σ_u for the material. A deep drawn cup may have to be completed in a series of operations.

Blank Size

In order to calculate the blank diameter D to produce a cup of diameter d, it is necessary to equate the surface area of each. This leads to the well known expression:

$$D = \sqrt{d^2 + 4dh} \quad \text{Where } h = \text{cup height}$$

This makes no allowance for a corner radius, and in practice one may finally adopt a trial and error technique; in drawing work experience is at a premium!

Force Required for Drawing

Again, the force will vary as the operation proceeds, but a reasonable approximation for the maximum punch force can be established using the expression:

$F_{max} = \pi . d . t . \sigma_u$ where $\sigma_u =$ the ultimate tensile stress of the material.

Example 12.4

A cup is to be drawn to a diameter of 76·2 mm × 38·0 mm deep in 0·8 mm thick material. Estimate the blank diameter and the maximum drawing force if $\sigma_u = 432$ N/mm^2

Solution

$$D = \sqrt{d^2 + 4dh}$$
$$= \sqrt{5\,800 + (4 \times 76 \cdot 2 \times 38 \cdot 0)} = \sqrt{5\,800 + 11\,600}$$
$$= \sqrt{17\,400} = 132 \text{ mm}$$

$$\text{Drawing ratio} = \frac{D}{d} = \frac{132}{76 \cdot 2} = 1 \cdot 73$$

∴ assume the cup can be drawn in one operation.

$$F_{max} = \pi . d . t . \sigma_u$$
$$= 3 \cdot 142 \times 76 \cdot 2 \times 0 \cdot 8 \times 0 \cdot 432 = 0 \cdot 083 \text{ MN}$$

12.5 COMBINATION TOOLS

It is often possible where batch quantities justify it to incorporate two operations into the one tool such that the operations are carried out in *combination*. This increases the production rate hence reducing the variable costs with little increase in fixed costs. The dual operation is usually blank and draw and is very successful on thin materials where the total force required is not impracticable, and the component can be drawn to full depth in the one operation. Figure 12.13 shows the principle of operation of such a tool.

The punch blanks upon diameter D and draws on diameter d. The punch and die pressure pads are spring loaded. As the punch descends in the guide, it shears the blank from the strip, and the blank will be gripped between the punch face and the pressure pads. The punch continues to descend, and the blank is drawn over the drawing die. The drawn cup is in an inverted position to that shown in Fig 12.11.

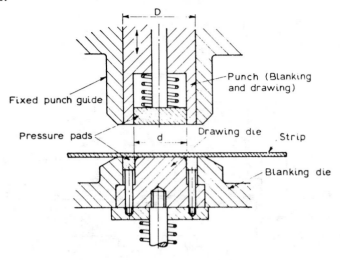

Fig 12.13   Combination Tool.

As the punch withdraws, the pressure pads act as ejectors so that the cup does not stick to the die or punch. The punch guide can also be arranged to act as a stripper, so that the strip is removed from the punch.

12.6 PROGRESSION TOOL

The combination tool described in the previous section combines more than one press operation at the one tool station, and the punches and dies are positioned on the same axis. A progression tool, by contrast, has more than one press operation positioned at separate stations on the press tool set, i.e., there is one station for each operation, these following on behind each other. The strip is therefore *progressively* worked upon as it is fed through the tool. Simple tools of this type, such as the tool shown at Fig 12.4 are called follow-on tools, and are usually blank and pierce tools.

Large progression tools can be very complex and expensive to make, therefore they are only suitable for high quantity production, and the strip feed will be automatic. This point has already been mentioned in Section 5.9 when discussing the merits of transfer pressing, which is an alternative to progression tools when large quantities of components are involved. Complex progression tools may include a combination type operation at one of the stations.

Figure 12.14 shows the principle of operation of a progression tool.

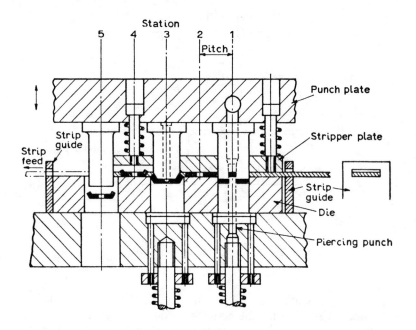

Fig 12.14 Progression Tool.

Assume the component being produced in the progression tool shown is some type of locking washer having formed tabs upon either side. The washer is blanked and pierced at station 1, and finished formed at station 3. In all, five stations are provided, spaced an equal pitch apart, the strip being roll fed through the guide and moving one pitch forward each time the punches withdraw. Stations 1, 3 and 5 are working stations, and stations 2 and 4 are idle stations. This is done to provide room for the punches, and to strengthen the die block. The blanks and formed washers are carried in the strip from station to station before the finished washer is finally pushed out of the strip by the ejector punch at station 5. Let us examine each operation at each station in turn.

Station 1
Blank and pierce operation. The fixed piercing punch in the bottom tool pierces a slug as the top tool descends. The slugs are pushed up the centre of the blanking punch to finally fall out through the exit hole in

the punch plate. The spring loaded pressure pad in the bottom tool supports the blank and pushes it back into the strip after blanking and piercing are complete.

Station 2
Idle operation. The stripper holds the blank and strip as the top tool descends.

Station 3
Forming operation. Bending of tabs takes place on the washer being supported by a pressure pad. Again, the pressure pad pushes the formed washer back into the strip to allow it to be carried to the next station. A pilot in the punch locates the pierced hole ensuring the blank is accurately positioned for forming.

Station 4
Idle operation. The stripper plate has a clearance slot cut in it so that the washer tabs are not damaged as the stripper plate grips the strip.

Station 5
Ejection operation. The finished washer is pushed out of the strip to fall clear of the tool into a hopper.

These tools are very fast in operation and the amount of scrap strip left to deal with can create problems. On some tools the scrap strip is cut up by the tool to be ejected off the die face using compressed air for example. Sometimes a jet of compressed air is used for ejection purposes upon a press tool.

12.7 RUBBER DIE FORMING

Press tools are a very efficient means of producing large quantities of sheet metal components, and well made tools may produce quantities of many thousands, up to 100 000 say. However, where small quantities such as 1 000, are required, a rubber press tool is a cheap and simple means of carrying out forming operations.

The example shown at Fig 12.15 is of a rubber press tool carrying out a bulging operation upon a canister.

The split die is assembled with the canister in place and the rubber cylindrical die placed inside the canister. The punch descends and forces a displacement of the rubber die which in turn forms the canister to the split die profile.

The operation is simple, but of course, comparatively slow. However, the operation shown is an ideal application, because this would be a difficult operation to complete by other means. It is estimated that rubber dies may have a life of from 8 000-20 000 parts, depending upon the severity of the operation. The process is usually used with soft, ductile materials.

Hydroforming

This is a newer technique than rubber forming and is considered to have great potential for the future. It uses a rubber diaphragm for displacing a sheet metal blank, but the rubber is backed up by high pressure oil.

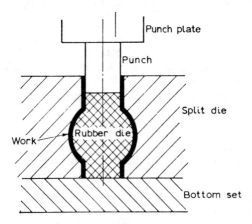

Fig 12.15 Rubber Press Tool.

12.8 HIGH ENERGY FORMING

Before completing this chapter we will briefly consider the new high energy forming processes. We have already considered the work done in pressing, and it should be noted that forming for example is normally carried out at the relatively slow punch speed of 0·10 - 0·25 m/s. High energy forming makes use of gas, air or explosives to provide greater energy for the pressing operation, the process being carried out at a speed much greater than normal. High energy forming has been successfully applied to both shearing and drawing. A one stroke petrol engine has been built into a press and has had good results in shearing at high speed. The sheared slug is found to have less deformation. Compressed air presses are available commercially for drawing operations, but require large compressors. They utilize high velocity high pressure air, and one version delivers 55 kJ of energy at 60 m/s. The principle of operation of such a press is shown at Fig 12.16.

At Fig 12.16(a) the compressed air is impinging on a small area, while the atmospheric pressure is impinging on the large area of the piston. This is in a state of equilibrium as shown, the piston being balanced by the pressures. As soon as the valve is closed [Fig 12.16(b)], the system becomes unbalanced and the piston descends at a very high speed. The air at atmospheric pressure exhausts as shown.

Other high energy mediums are being experimented with, such as high voltage discharge, liquid nitrogen and also explosives.

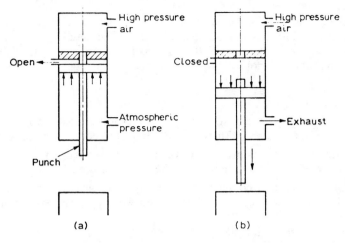

Fig 12.16 High Energy Forming Press.

12.9 EXPLOSIVE FORMING

This process is in its infancy, but appears to have such potential that it may well prove to be an important forming process in future years. It will probably find an application where large, difficult forms are required in small quantities; the process is slow at its present stage of development. A typical operation would be that of bulging as shown in Fig 12.15. A die would be used, but an explosive charge would be suspended in the work to bulge out the canister, instead of the rubber shown. Hence, no punch is required. The operation can be carried out in the open (at a safe distance!) or more generally with the dies immersed in water. This stand-off explosive forming, as it is called, utilizes the shock waves in the water instead of in the air to deform the component when the explosion takes place. The explosive generates terrific power causing large hydrostatic forces, and plastic deformation is achieved by mechanical means. So far most of the experimental work has been confined to work which can be formed and extracted in a solid die. Epoxy resin dies have been successfully used, because of the recovery properties of the material.

Explosives such as I.C.I. Cordex can be used in lengths, this developing 5·4 kJ of energy/m, one metre weighing 280 g. Powder explosive can be used when charged in a plastic sleeve. Small electrical detonators ($\frac{1}{2}$ amp) are required to ignite the explosive charge. The operation speed can be measured in micro-seconds. If accurate pressings are required, the operation should be carried out with the air in the die evacuated.

If air is present between the blank and the die walls, then this trapped air can cause distortion of the work.

At present, the disadvantages of this process are the slow operating rate, the obvious dangers which are present when handling any type of explosive (this danger will always make the process a specialized activity), and the noise from the operation.

Exercises 12

1. By considering the costs involved, compare the cold working of a component to shape, with the machining of a component to shape.

2. Discuss the factors involved in the cold pressworking processes and show the effect of these factors on the design of press tools.

3. (a) Contrast and compare with the aid of diagrams, a mechanical crank press and a hydraulic press. For what purpose is each most suited?

(b) Show the principle of operation of a press guard for a small capacity press engaged upon a blanking operation, the press being operator controlled.

(c) What is a double acting press, and for what purpose(s) would it be used?

4. A steel blank 45 mm square having a 22 mm diameter hole in the centre is to be blanked from 1 mm thick sheet. (a) Calculate the maximum punch force necessary to shear the blank in one operation if $\tau = 390$ N/mm².

(b) Calculate the work done if the % penetration is 20%.

(c) What will be the % reduction in punch force if 0·5 mm double shear is ground upon the tool.
$$\text{(Ans. (a) } 0\cdot097 \text{ MN, (b) } 19\cdot4 \text{ J, (c) } 71\cdot3\%)$$

5. A component is formed on a bending tool from 6 mm × 35 mm material into a 'U' shape 55 mm wide, having unequal arms 62 mm and 48 mm long respectively. The internal radii at each bend corner is 12 mm. Calculate (a) the maximum punch force for bending the part in one operation.

(b) The additional force if planishing is carried out, and

(c) The developed length of the part (ignoring bend allowances) Take τ as 400 N/mm² and σ_y as 350 N/mm².
$$\text{(Ans. (a) } 0\cdot084 \text{ MN, (b) } 0\cdot59 \text{ MN, (c) } 140\cdot2 \text{ mm)}$$

6. Estimate (a) the maximum punch force, and (b) the blank diameter, for a cup which is to be drawn to 62·3 mm dia. × 21·4 mm deep in 0·8 mm thick material. $\sigma_u = 325$ N/mm²
$$\text{(Ans. (a) } 0\cdot051 \text{ MN, (b) } 96 \text{ mm)}$$

7. Clearly show the difference between combination tools and progression tools, and sketch an example of each type. What is the fundamental difference between transfer pressing and progression tool pressing?

8. Explain the principle of operation of rubber forming, and hydro-forming, as applied to presswork. Show an example of one kind of tool.

9. What is high energy forming, and in what form is the principle applied? Sketch an example of one application.

10. A 20 mm diameter vee pulley of very simple design for use on a washing machine, can be produced in large quantitites, either by (a) pressing and fabricating in two parts from sheet metal, or (b) machining upon an automatic from aluminium bar stock. Design a pulley and sketch the tooling required for each method of manufacture. Using a break-even analysis and your own cost information, find the break-even quantity.

Further Reading

1) Noble C. F. 'Short-run Sheet Metal Production Methods.' *Journal of the Institute of Mechanical Engineers*, July 1974.

2) Lissaman A. J. and Martin S. J. *Principles of Engineering Production.* Hodder and Stoughton Ltd.

3) Hinman S. *Pressworking of Metals.* McGraw-Hill Book Co. Inc.

4) Willis J. *Deep Drawing.* Butterworths Scientific Publications.

5) Galloway D. F. 'Modern Trends in the Manipulation of Metals.' *Journal of Institute of Production Engineers*, 1960.

6) Hollis W. S. 'Advanced Methods for Forming Metals.' *Engineering*, July 1960.

7) Hollis W. S. 'The Process of Explosive Forming of Metal.' *Journal of Institute of Production Engineers*, 1960.

8) Martin A. E. 'Bending, Forming and Shearing.' *Journal of Institute of Mechanical Engineers*. Jan. 1970.

9) Ford H. 'Forming and Shaping Operations for Iron and Steel'. *Journal of the Institute of Mechanical Engineers*, April 1973.

10) Sawle R. 'Forming in Superplastic aluminium.' *Journal of the Institute of Mechanical Engineers*, Dec. 1978.

CHAPTER 13

Specification of Quality and Reliability

13.1 QUALITY

QUALITY IN its broadest sense, means *degree of excellence*. In the last century, people of aristocratic blood were termed people of quality, as opposed to the great mass of ordinary people. We have all compared a *good quality* product with an inferior one. The former looks good, is adequately strong, functions efficiently and reliably, and is well made. Is the quality a result of the design, or a result of the manufacture? Well, of course, there is some of each present, and we can say that there are two factors which have an effect upon quality:

a) Design — quality can be specified by a designer on his drawing.

b) Measurement — quality can be measured and controlled on the shop floor. We will consider this factor of quality control in Section 13.9 and Chapter 14.

The other thing about our good quality product is that it costs more than the inferior model, but we think this state of affairs is satisfactory because we get reliability and a long product life for our money. However, if it lasts too long and becomes obsolete because fashions change, then perhaps we might think we paid a little too much for it. Quality then, is connected with cost, and the two must be balanced to suit the particular class of people to whom we are hoping to sell the product. When we refer to quality in this chapter we shall be referring to those characteristics of a product specified at the design stage, such as degree of finish, material and shape, types of fit, and degree of accuracy of dimension; remembering that the whole is also a question of economics.

Elsewhere (see Further Reading, No. 1) it has been stated that: 'the modern conception of the creation and control of quality, by definition, means the development and realization of specifications necessary to produce economically, and in adequate degree, the appearance, efficiency, interchangeability and life which will ensure the product's present and future market'. This definition emphasizes the creation of quality by design, and by measurement. It also draws our attention to the importance of a *specification*, if we wish to economically manu-

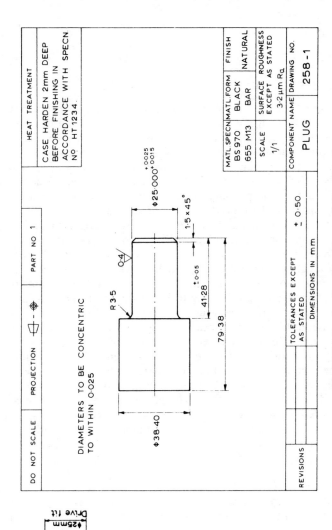

Fig 13.1 Component Drawing.

Fig 13.2 Production Drawing and Specification.

facture a product, having the quality level we desire, to suit the market.

13.2 SPECIFICATION

A specification is a detailed description, and in order for the description to be *specific* it must be *precise*. A specification can be drawn up after the design drawings have been completed, and will be necessary before manufacture can take place. Basically, the designer's first considerations are that the product should be functional and very often that it should also have aesthetic appeal. (Note. An object which has aesthetic appeal is beautiful and pleasing to look at, this often being implicit in the term *styling*.) On the other hand, the production engineer's first consideration is that the method of manufacture shall be efficient and economically sound; he is concerned with men, machines, material and money. There is sometimes a conflict between the designer's aims and the production engineer's aims, which we will refer to in more detail in the next section. A specification should enable a part shown upon an engineering design drawing to be produced on the shop floor with all the required features accurately reproduced.

Let us take a simple example. Figure 13.1 shows a component which is required by the designer in the form shown.

This drawing does not contain nearly enough information to enable the part to be produced in quantity to the precise quality level required. Therefore, the production engineer must produce a working drawing and specification which is without ambiguity in order to achieve the desired result. It is important that everyone concerned in the manufacturing organisation uses the same standards in order that the language of the specification is understood by all. We shall discuss standardization in Section 13.4, but it is not feasible in a book of this type to discuss standards in detail. We will refer to some where appropriate, and the reader is urged to look at actual copies and to regard it as part of his education to understand them. Returning to the drawing shown at Fig 13.1, BS 308 (Engineering Drawing Office Practice) should be used in order that engineering drawing conventions may be standardized, such as dimensions, projections, sections, etc. Also, decimal dimensions, not fractional, should be used. For example the $3\frac{1}{2}$ mm dimension shown in Fig 13.1 would be represented by 3·5 mm or 3·500 mm.

The accepted convention is as follows: 3·5 mm implies an accuracy of 0·1 mm; 3·50 mm implies an accuracy of 0·01 mm; 3·500 mm implies an accuracy of 0·001 mm. Dimensions are normally given in mm units, not cm.

Figure 13.2 shows a production drawing and specification for the part shown at Fig 13.1.

In this simple example, the working drawing and specification are shown together. It will be appreciated that for more complex components, the specification might have to be a separate document.

Referring to Fig 13.2 it can be seen that the factors affecting quality have now been more precisely specified. These are:

a) *Material*. Instead of steel, a particular material specification has been referred to, viz., BS 970, 655M13. This is the standard for wrought steels in the form of bars, billets and forgings. It lays down the chemical composition of the steel to be used, for example, 655M13 is a $0.1/0.16\%$ C, $3\frac{1}{4}\%$ Ni, case hardening steel. It also specifies the properties which the steel must possess after heat treatment, for example, 655M13 must possess a tensile strength of 1 000 N/mm^2 minimum, Izod value of 40 J minimum, etc.

b) *Surface finish*. The degree of surface roughness is specified by the symbol $\overset{0.8}{\diagdown\!\!\diagup}$ for example. The current method of indicating surface roughness is shown in BS 308. The meaning of the symbol and the principles underlying surface finish measurement can be found in BS 1134 (Assessment of surface texture).

This specification of surface finish leads the production engineer to the most appropriate process which can be utilized. A surface finish value of 0.4 μmR_a for example can easily be attained by grinding, but would be outside the scope of a production turning process. Therefore the 25·000 mm diameter on the part shown at Fig 13.2 will require grinding after hardening, this process also being appropriate for the dimensional tolerance of 0·010 mm which is specified.

Any type of finish such as electroplating, painting, etc., which is required, should be specified in addition to the machining finish. If not required then the finish is specified as 'natural'.

c) *Dimensions*. Precise dimensional limits have been specified for the 25 mm diameter drive fit required by the designer. The 41·28 mm length is quite important, and appropriate limits have been specified. Note that the imposition of unnecessarily tight limits increases the cost of the product. All the other 'open' dimensions have had a general tolerance applied. Even relatively unimportant features such as the 15 mm chamfer must be specified.

The decimal system of dimensioning referred to earlier has been adopted in the interests of standard practice.

The 38·1 mm length has been omitted because the 41·28 mm length and the 79·38 mm overall length are important. In effect, the datum face is the right hand end face of the part as drawn, and these two lengths are specified from this datum. Therefore, the 38·1 mm length is unimportant and is not specified. The subject of standards for limits and fits is discussed in Section 13.6.

d) *Geometric relationships*. If the concentricity of diameters relative to each other is important, then this must be specified, as must any other geometric relationship of the component features. This could, and often does, affect the method of manufacture. The permitted amount of tolerance for the geometrical relationships must be specified as shown in Fig 13.2, in the same way as dimensional tolerances are specified. These geometrical relationships are outlined in BS 308, and this important point will be dealt with more fully in Section 13.8.

13.3 DESIGNING FOR PRODUCTION

In Section 13.1 we referred to a conflict which can often arise in a manufacturing organization between the designer's aims and the production engineer's aims. We have tried to show that if manufacture is to proceed efficiently and at minimum cost, then a specification is necessary. An ideal situation would arise if the designer was capable of creating a design of a product which was satisfactory from the aesthetic and functional viewpoints, and also ideal from the economic and production viewpoint. This total conception of design (with the customer, quality, cost and manufacture equally considered) has lead to the widespread use in industry of the maxim 'design for production'.

It is generally thought unrealistic to expect designers to be specialists both in the functional aspects and the production engineering aspects. Therefore the ideal arrangement is to have a link man or liaison engineer in the organisation who works closely with both sides. This engineer will be primarily a production engineer well experienced in manufacturing technology, capable of modifying designs in order to make their production more efficient, but still allowing the designer to retain the essential functional features of the design. This is part of the methods function of a work study department, and in some organizations the liaison man is called a product methods engineer. Whatever the name, this activity is one of the most important in the cycle of a product from its inception to its completion. The product methods engineer should work on the project at the design stage, collaborate in drawing up a specification, advise at the planning stage and during the design, manufacture, layout and installation of tools and equipment, and also during actual manufacture and inspection. This routine will obviously vary depending upon the size of the firm and the project, but the accent must always be on team work and close collaboration between the engineering and design staff, and the production staff. Unhappily, the opposite is sometimes evident, the sufferers being the firm and ultimately the firm's employees.

13.4 STANDARDIZATION

In Section 5.1 we discussed the concept of standardization, but we must consider it again as it is closely bound up with the task of preparing specifications.

Standardization
This is a process which leads to the establishment of desirable criteria with respect to material, size, etc., to which everyone can adhere. If

they do, then they would expect that the standard would meet all their requirements and enable them to manufacture and stock the minimum variety of parts. The greatest economic benefit can be obtained if every design makes use of standard parts, such as screws, splines, gears, etc., manufactured with standard cutting tools, inspected using standard gauges, stored and despatched in standard containers. This sounds as if all engineering products are to be dull and uniform objects conforming to standardized objectives. However, the skilled designer can still be creative and stimulating whilst making full use of standard materials and parts. Every time he calls for a special item, the work of specification preparation is made more difficult and costs rise.

Standards are available to the engineer to cover dimensions and tolerances, material specification and properties, machines, tools and gauges, methods of assessing and inspecting and installation and rating. The application of standards affects quality, and if the criteria which have been established in the standard are desirable, then the quality level should be enhanced. The first subject mentioned above of dimensions and tolerances is to do with the physical size and type of fit of mating parts (Section 13.6). This facet of standardization makes possible the vitally important system of interchangeability.

Interchangeability

This is a manufacturing system in which every one of every batch of parts produced will assemble freely with any one of all the mating parts produced, i.e., every mating part is interchangeable with every other, without any fitting or modification to the parts being necessary. If different firms are producing the parts to the same standard specification, then all of the firms' separate products will be interchangeable. This reduces costs because the task of assembly is simplified, and also standard replacement parts can be drawn from stock in the certain knowledge that they will fit without alteration.

In order to achieve interchangeability appropriate component tolerances must be specified from the standard to suit the type of fit required. Then the chosen manufacturing process must have the correct degree of precision to be capable of producing within the required tolerances. Finally, the system of quality control used on the shop floor must ensure that only components within the specification are accepted for use.

Simplification

This concept is closely related to standardization, and because of it a company's range or variety of products can be reduced, i.e., simplification is a process of variety reduction. Standardization means that the manufacturing effort can be concentrated upon a narrower front, and parts produced with less effort and skill being required. This leads to simplification and less variety of products, but production should be at minimum costs as a result with a consequent advantage in sales price.

Again we have a source of conflict between the production engineer and in this case the sales department. The former will want standardization and simplification to be applied as widely as possible for the reasons previously stated. To sum up his case, the less the variety, the longer the production runs, and the less the number of set ups and tool changes. On the other hand, the salesman wants greater variety of product in order to offer the customer the widest possible choice in order to improve sales. The ideal position is somewhere in between both extremes, where unnecessary variety is deleted, but a sufficiently wide range of products exist to attract a fair proportion of market demand. Market research is a technique which can provide valuable information in this situation, if economically justified.

Specialization
This is the natural outcome of the application of standardization and simplification. In this context we do not mean the specialization of effort as applied to human activities on the shop floor (Section 5.1), but rather the specialized activities of firms concentrating their productive effort upon a limited number of products. Because of standardizing and simplifying, they will become more expert and specialistic at producing a certain range of products, and enjoy all the economic advantages which accrue.

13.5 PREFERRED NUMBERS

Preferred numbers are a series of numbers which are selected for the purpose of standardization and simplification, in preference to other numbers. The idea of preferred numbers was first introduced by Colonel Charles Renard in 1870, and is sometimes referred to as the Renard series. He was faced with a simplification problem because 425 different sizes of cables were used in his army unit. In carrying out a process of standardization in order to reduce the range of sizes, he suggested that a geometric series should be used as a basis of selection. His series is shown below, the series being denoted by an 'R'.

Renard Series of Preferred Numbers

Series	Ratio	Steps increase by
R5	$\sqrt[5]{10} = 1{\cdot}58$	60%
R10	$\sqrt[10]{10} = 1{\cdot}26$	25%
R20	$\sqrt[20]{10} = 1{\cdot}12$	12%
R40	$\sqrt[40]{10} = 1{\cdot}06$	6%
R80	$\sqrt[80]{10} = 1{\cdot}03$	3%

The standardizing organizations such as B.S.I. (British Standards Institution), I.S.O. (International Organisation for Standardization) and A.S.A. (American Standards Association), have issued standards based upon the 'R' series. The reader is referred to BS 2045 (Preferred

Numbers). Many standard size ranges for products such as sheet steel, drawn wire, bolts, etc., will be found to follow a geometric series.

Let us take an example to illustrate the use of preferred numbers.

Example 13.1

A manufacturer wishes to produce a range of refrigerators. The range has to include six models, the smallest to have a cabinet capacity of 0.06 m³. Calculate the range of sizes if they are to rise in geometric progression.

Solution

The R5 series can be used, this being a geometric progression of the form:

a, ar, ar^2, ar^3, ar^4, and ar^5 having six terms and five steps, where a is the first term, and r is the rate of increase of the steps.

$$\text{In the R5 series, } r = \sqrt[5]{10} = 1.585$$

∴ the range of sizes will be:

First size $\quad= 0.06$ m³ (given)
Second size $= 0.06 \times 1.585 \quad= 0.095$, say 0.10 m³
Third size $\quad= 0.06 \times (1.585)^2 = 0.15$ m³
Fourth size $= 0.06 \times (1.585)^3 = 0.239$, say 0.24 m³
Fifth size $\quad= 0.06 \times (1.585)^4 = 0.378$, say 0.38 m³
Sixth size $\quad= 0.06 \times (1.585)^5 = 0.6$ m³

If the R10 series were used, then the range of sizes would start at 0.06 m³ rising in 10 steps to 0.60 m³. The R80 series gives 80 steps, such a large range as this rarely being used in practice. Of course, only part of a series may be used if required.

Economic Aspects of Preferred Sizes

The justification for using preferred numbers in a programme of standardization and simplification is that unnecessary overlapping of sizes is eliminated. In Example 13.1, the product sizes rise in steps of approximately 60%, and it would appear to add unnecessary variety to the range to include yet a further size of 0.13 m³ say, in that 0.13 overlaps the 0.1 and 0.15 sizes. The addition of such an extra model could be economically unsound, because high production of a narrow range of products can be carried out more cheaply than the production of a wider range of products.

There is, however, another side to this argument. Imagine that the refrigerator manufacturer of Example 13.1 already has an existing range of seven models; the R5 series plus the 0.13 model. He wishes to simplify the range in order to reduce his manufacturing costs, and probably reduce his prices. If the preferred series becomes the criterion, then the 0.13 model must be eliminated. However, this may be the most popular model in the range to the customer, and the most profitable model to the firm; therefore to eliminate it from the range would appear to be wrong. Again, it may be that the wholesaler has an occasional demand from the retailers for a 'special 0·13' model, which cannot be obtained from any other source. Therefore, the wholesaler takes a good proportion of the standard models (which may not be the best on the market) from our manufacturer, in order to keep up the supply of the 0·13 model which he cannot otherwise obtain.

Economic considerations are rarely straightforward and simple, and it is therefore imperative that a programme of simplification is carried out using accurate and careful cost analysis. Valuable information can be obtained using a break-even chart (Section 1.3) for each product in the range, plotting total sales income (£) against total costs (£). This establishes the break-even point between profit and loss for each model.

13.6 LIMITS AND FITS

Interchangeable manufacture is only possible if two factors are established:
1) The type of fit which is required between mating parts, such as running fit, push fit, drive fit, etc.
2) The tolerance which is to be allowed upon each dimension.

The designer can estimate these, hence creating special conditions and increasing cost, or appropriate standards can be specified, giving the benefits we have previously discussed. The two best known standard systems of limits and fits are BS 4500: 1969 (ISO limits and fits in metric units) and the Newall Limit system, prepared by the Newall Engineering Co. Ltd. The former is comprehensive and is the most important of the limit systems, the latter is simpler and has probably been used far less since BS 1916 (limits and fits in imperial units) was published in its present form in 1953. Some large manufacturing units produce many of their own standards.

Consider Fig 13.3 which shows the basic principles underlying any systems of limits and fits.

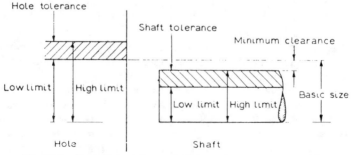

Fig 13.3   Basic Principles underlying Limits and Fits System.

Figure 13.3 shows the following factors:

a) *Basic Size.* This is the nominal size specified by the designer, but which cannot be produced exactly because of the inherent inaccuracies of manufacturing processes.

b) *Tolerance.* This is the amount of variation which can be tolerated to allow for the inherent inaccuracies of the manufacturing process. It is the difference between high limit and low limit.

c) *Limits.* The high limit is the largest permissible dimension, and the low limit is the smallest permissible dimension.

d) *Clearance.* The shaft shown is a clearance fit, this amount of clearance being affected by the chosen tolerance band.

Example 13.2

In a limit system, the following limits are specified to give a clearance fit between a hole and shaft of 25 mm nominal diameter.

hole	25 mm diameter
	+ 0·020
shaft	25 mm diameter
	− 0·005
	− 0·018

Determine (a) the basic size, (b) the hole and shaft tolerances, (c) the hole and shaft limits, (d) the minimum and maximum clearance.

Solution

a) Basic size = 25 mm
b) Hole tolerance = 0·020 mm
   Shaft tolerance = 0·013 mm
c) Hole high limit = 25·020 mm
   Hole low limit = 25·000 mm
   Shaft high limit = 24·995 mm
   Shaft low limit = 24·982 mm
d) Minimum clearance = Hole low limit − shaft high limit
   = 25·000 − 24·995 = 0·005 mm
   Maximum clearance = Hole high limit − shaft low limit
   = 25·020 − 24·982 = 0·038 mm.

Types of Fit

Example 13.2 corresponds to Fig 13.3 in that a *clearance fit* is referred to, i.e., the largest possible shaft is less in diameter than the smallest possible hole. There are two other types of fit.

An *interference fit* results from the smallest possible shaft being greater in diameter than the largest possible hole.

A *transition* (in between two conditions) *fit* results from a condition in which tolerances are disposed in such a way that the fit may be clearance or interference, e.g., the smallest shaft could be clearance fit with the smallest hole, and the largest shaft could be an interference fit with the largest hole.

These fits are illustrated in Fig 13.4.

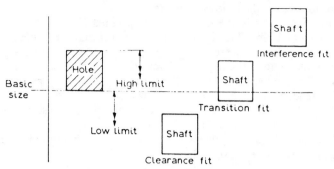

Fig 13.4 Types of Fits.

Figure 13.4 shows the accepted way of representing limits and fits diagrammatically, only a representation of the tolerance and its disposition being necessary for each mating part. The fits shown at Fig 13.4 are achieved by maintaining the hole at a constant size and varying the size of the shaft. This is known as a *hole basis* system, and all modern systems are based upon this principle. This is because most holes machined by quantity production methods are held to some standard size, using a fixed diameter cutter, such as a drill or reamer, for example. It is therefore more economic to specify a change in shaft diameter to give the required type of fit.

Tolerance Disposition

If the tolerance is shown disposed on one (uni) side of the basic size, this is called a *unilateral tolerance*. For example:

$$31 \cdot 500 \text{ mm} + 0 \cdot 002 \quad \text{or} \quad 31 \cdot 500 \text{ mm} - 0 \cdot 003$$
$$+ 0 \cdot 001 \qquad\qquad\qquad - 0 \cdot 001$$

The clearance fit shaft shown in Fig 13.4 has a unilateral tolerance, as also has the hole.

If the tolerance is shown disposed on both (bi) sides of the basic size, this is called *bilateral tolerance*. For example:

$$31 \cdot 500 \text{ mm} + 0 \cdot 004$$
$$- 0 \cdot 004$$

The transition fit shaft shown in Fig 13.4 has a bilateral tolerance.

Unilateral tolerances are preferred in modern limit systems, because in a hole based system the basic size will always be the GO size of a limit plug gauge, which is very convenient and practical (see Fig. 13.4).

Fundamental Deviation (FD)

The deviation of the tolerance band (on shaft or hole) away from the basic size is called the fundamental deviation. Therefore, the hole FD will be the difference between the basic size and the hole low limit; the shaft FD will be the difference between the basic size and the shaft high limit. In Fig 13.3 the hole FD = 0, and the shaft FD = the minimum clearance. In Fig 13.4 the hole FD = 0, and the interference fit shaft

FD = the maximum interference. It can be seen then that the magnitude of the tolerance, and its position relative to the basic size as determined by the FD, will determine the class of fit between hole and shaft.

Newall Limit System (at the time of writing available in Imperial units only)

This is a bilateral, hole basis system for sizes up to 150 mm. It has two classes of hole (A and B), and six classes of shaft (F, D, P, X, Y, Z) combinations of which give force fit, driving fit, push fit or three grades of running fit. It is a simple system, and once the basic size is known, the limits can be taken from the table for any one of the grades of fit. For example:

A class A hole combined with a class F shaft of 25 mm basic size gives the following limits:

Hole	*Shaft*
25·000 mm	25·000 mm
dia.	dia.
+ 0·013	+ 0·050
− 0·006	+ 0·038

This gives a maximum interference of 0·056 mm and a minimum interference of 0·025 mm. This grade of force fit would require the mating parts to be shrunk together, or hydraulically pressed together

BS 4500:1969

This is a hole basis or shaft basis limit system which covers a size range up to 3 150 mm. Therefore, engineering organisations would generally only find themselves concerned with a particular part of the standard to suit their requirements. We stated earlier that the class of fit was determined by the magnitude of the tolerance and the FD. Let us briefly examine how these are arranged in BS 4500.

Tolerance

If a particular tolerance gives certain functional conditions of the mating parts for a smaller basic size, the same tolerance imposed upon a larger size will give different functional conditions For example, a tolerance of 0·36 mm upon a basic size of 76 mm gives certain operating conditions at a certain production cost. In order to maintain the same operating conditions at the same cost on a basic size of 380 mm, the tolerances must be increased, say to 0·635 mm. This is shown diagrammatically at Fig 13.5, and the example cited giving a constant quality set of conditions over the whole size range would be called a tolerance grade.

To maintain the constant quality conditions (with respect to size only) as described above, the tolerances are increased in steps (shown dotted Fig 13.5) as the basic size increases. BS 4500 has eighteen tolerance grades from which to choose, these being designated IT 01, IT 0, IT 1, IT 2 etc., up to IT 16.

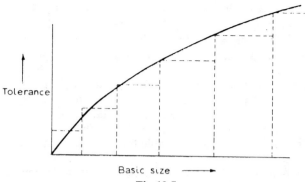

Fig 13.5

Fundamental Deviation

The FD in BS 4500 is designated by single or pairs of capital letters for holes, i.e., A to ZC, and similarly by small letters for shafts, i.e., a to zc. There are 41 FD's of differing magnitude specified. In general the FD for a shaft, and the FD for a hole of the same letter and basic size are equal. For example, a 25 mm hole having limits of $+0.013$ and a 25 mm $+0.000$
mating shaft having limits of -0.013 both have fundamental deviations -0.000
equal to zero.

Note that the *maximum metal condition,* (denoted by MMC, which means largest possible shaft or smallest possible hole) is equal to the basic size for both shaft and hole.

Class of Fit

The wide range of tolerances and fundamental deviations which can be selected gives many classes of fit, although some of the possible combinations which can be derived may not be of practical use.

When certain shafts and holes of a particular fundamental deviation and tolerance grade are selected, they should be specified in this manner:

    (*shafts*)　25 mm s6
    (*holes*)　25 mm H7

This means that the H7 hole (of H designation fundamental deviation, and 7 grade tolerance) and the s6 shaft (of s designation fundamental deviation, and 6 grade tolerance) give the following limits when combined.

Hole	Shaft
25·000 mm dia.	25·000 mm dia.
+0·021	+0·048
	+0·035

This gives a maximum interference of 0·048 mm, and a minimum interference of 0·014 mm.

13.7 TOLERANCE BUILD-UP

Consider the part shown at Fig 13.1 and again at Fig 13.2. The variation of the 38·1 mm length will depend upon how much the 41·28 mm ± 0·05 and the 79·38 mm ± 0·50 lengths vary. It will be at its largest possible value when the 79·38 dimension is at its largest (i.e., 79·88 mm), and the 41·28 dimension is at its smallest (i.e., 41·23). Therefore, the argest possible value of the nominal length of 38·1 mm is 38·65 mm. By similar argument its smallest possible value is 37·55 mm. The difference between these two limiting dimensions is 38·65 − 37·55 = 1·10 mm, which is the greatest possible variation of the 38·1 mm dimension. Note that this is equal to the sum of the tolerances of all the other dimensions of length, i.e., 0·10 mm + 1·00 mm.

The maximum variation of this dimension above the nominal size of 38·1 mm is 0·55 mm, and the maximum variation below the nominal size is 0·55 mm. Now suppose that the 38·1 mm dimension is important and is specified as 38·1 mm + 0·18 having a tolerance of 0·13 mm,
+ 0·05
and the overall length is unimportant. Therefore the build-up of tolerances over the whole length will now be 0·13 + 0·10 = 0·23 mm, and the overall length of 79·38 mm cannot now vary by more than this amount. Its largest value will be 41·33 + 38·28 = 79·61 mm, and its smallest value will be 41·23 + 38·15 = 79·38 mm the difference being 0·23 mm. For the purpose of manufacture, this overall length dimension is superfluous on the drawing, but if necessary it may be given as an *auxiliary* dimension as shown at Fig 13.6, its mean value being stated in brackets.

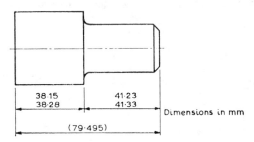

38·15
38·28

41·23
41·33

Dimensions in mm

(79·495)

Fig. 13.6

The dimensions are indicated in a different manner in Fig 13.6 than in Fig 13.2, the limits being shown in a different, but alternative fashion. Limits specified thus give no indication of the basic size, but the high and low limits are apparent at a glance, without further calculation being necessary.

Auxiliary dimensions are sometimes shown in a specification although they may not be required for the purposes of manufacture on the shop floor. They may however, be necessary for the design of tools or gauges.

Tolerance build-up then, must be taken into consideration during specification, otherwise it is possible to have conflicting requirements upon a drawing giving difficulties during the manufacture planning stage, leading to unsatisfactory functioning of the finished product.

13.8 GEOMETRIC TOLERANCES

In the last two sections we have been concerned with linear dimensions. Also the geometric features of a component such as straightness, flatness, etc., must be considered and specified, where these features are important and must be controlled. Geometric tolerances are concerned with the accuracy of the relationship of one feature to another, and now, where necessary, it is accepted that separate tolerances should be specified for geometric features, in addition to linear tolerances. In the past this factor was often glossed over, and it was considered that a diameter tolerance would also include an allowance for straightness, say. It is now generally agreed that this is not good enough. Linear tolerances and geometric tolerances when laid down separately in a specification prevent ambiguity. The principles of geometric tolerancing have shown the importance of the maximum metal condition (MMC) which refers to the condition of a hole or shaft when the maximum amount of material is left on, i.e., high limit shaft and low limit hole. It is always these limits which critically affect interchangeability of parts.

BS 308 gives an interpretation of geometrical tolerances for straightness, flatness, parallelism, squareness, angularity, concentricity, symmetry and position.

To illustrate the principle of geometric tolerances, consider the cylinder shown at Fig 13.7.

Fig 13.7   Plain Cylinder.

Firstly, a diametral linear tolerance of 0·076 mm is specified (we are ignoring the cylinder length in order to simplify the example). This means that the cylinder diameter, when measured at any point along its length, must be of some value between 30·480 mm and 30·556 mm.

Secondly, a geometric tolerance of 0·05 mm is specified for straightness. This geometric tolerance zone specified can be interpreted as shown in Fig 13.8.

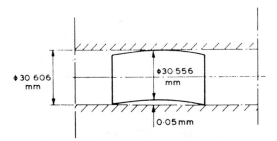

Fig 13.8   Geometric Tolerance.

As inferred earlier the worst condition for assembly of the cylinder into its mating part occurs when the mating features are in the maximum metal condition, and also if in addition the maximum errors permitted by the geometric tolerance are present. The cylinder is in this worst dimensional situation as depicted in Fig 13.8, and will just enter a truly cylindrical hole of diameter equal to the cylinder diameter plus both the linear and geometric tolerances, i.e., 30·606 mm diameter. Obviously, if all the linear tolerance is not taken up and the cylinder is 30·48 mm diameter for example, then there is a greater straightness tolerance available without endangering free assembly.

The measurement of the linear and geometric features discussed is shown diagrammatically in Fig 13.9.

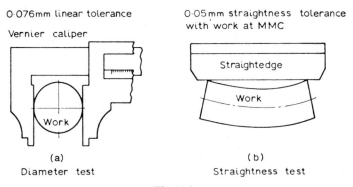

Fig 13.9

When considering more difficult examples, it is clear that there is much room for ambiguity and misunderstanding of the meaning of geometric tolerances. This fact only emphasizes the importance of building up a specification from the established standard, BS 308, if confusion is to be avoided.

13.9 LIMIT GAUGING

In order to ensure that the specification is being complied with on the shop floor, it is necessary to introduce some form of inspection. This may be carried out by *measurement* or *limit gauging* (usually called gauging). The differences between these two methods were discussed in Section 9.1, and we can now add a little to that. Figure 13.9 shows that the measurement of both linear and geometric features can be long, requiring some skill. On the other hand, a properly designed limit gauge checks both linear and geometric features simultaneously and can easily be used by semi-skilled operatives. Therefore, gauging systems of inspection are found in use most frequently in manufacturing organizations. Gauges may vary from simple hand gauges, to complex receiver gauges or gauging fixtures in which the work is presented to the gauge. We shall only have space in this section to consider the basic principles of gauge design.

Fig 13.10 shows a sketch of a simple *plug gauge* commonly used for the checking of a machined bore.

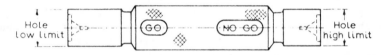

Gauging members are hardened and ground

Fig 13.10   Plug Gauge.

These gauges are made to many different designs, but in each case the gauge GO end is made to the component bore low limit of diameter, and the gauge NOT GO end is made to the bore high limit. Therefore the GO gauging member should enter the bore of an acceptable component, and the NOT GO gauging member should not enter the bore, hence it is much shorter in length. Also, the GO gauge should be truly round and straight within the specified geometric tolerances in order that the geometric features of the bore may be checked. It should be remembered, however, that the gauging members themselves must have a manufacturing tolerance, and will also be subject to wear. This point will be dealt with later in this section.

Taylor's Principle

This important gauging principle, first formulated in 1905, states that the GO gauge should be of full form (i.e., it should check both geometric features and size), while the NOT GO gauge should check only one linear dimension. The truth of this principle can best be understood if one looks at an example of its application. Let us take the case of a rectangular hole, with a tolerance zone as shown in Fig 13.11, which is to be gauged for both linear and geometric features.

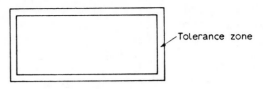

Hole to be gauged

Fig 13.11

Let us first see how the principle works if there is an error of geometry, say for example the corners of the rectangle are not square. Only a full form GO gauge will reject the work for this error. Simple linear gauges of the pin gauge pattern, made to the hole low limits, will not detect this error, as it is possible that they will still enter the hole, indicating the hole is satisfactory. This condition is shown at Fig 13.12.

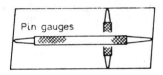

Hole having geometric error

Fig 13.12

Secondly, consider a linear error, or error of size; say for example the hole length is too great being outside the high limit. Only a NOT GO gauge which checks one linear dimension will reject the work for this error. A full form NOT GO gauge made to the hole high limits will not detect this error, as the gauge does not enter the hole because the width is within limits, although the length is outside limits. Again, such a gauge is indicating that the hole is satisfactory when it is not. This condition is shown at Fig 13.13.

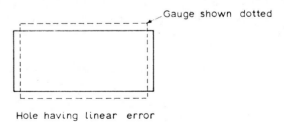

Hole having linear error

Fig 13.13

Correctly designed gauges conforming to Taylor's principle will consist of a full form, rectangular GO gauge, and two NOT GO pin gauges for the whole width and length. These gauges are used to test the hole surface in several positions in order to detect possible errors.

Sometimes, gauges are not designed to the principle outlined above, where geometric features are not important, or to make the gauge design and manufacture easier and cheaper. By now the reader will be aware that the plug gauge shown at Fig 13.10, although in common usage, does not comply to Taylor's principle. In order to do so, the NOT GO end would have to be modified as shown in Fig 13.14.

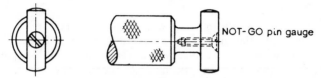

Fig 13.14   Hole Gauging Member.

Let us take one last example of gauge design relating to the checking of external round work, such as that shown at Fig 13.7. A common type of gauge used for this purpose is a caliper or *snap gauge* of the type shown at Fig 13.15.

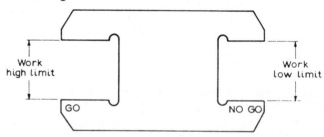

Fig 13.15   Caliper Gauge.

In order for the gauging to be correct in principle, the GO gauge should be of full form, which is not so in the caliper gauge shown Therefore, a single ended NOT GO caliper gauge is required, plus a GO ring gauge of the type shown in Fig 13.16.

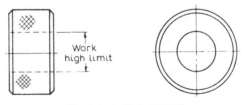

Fig 13.16   Ring Gauge.

Gauge Tolerances

Ideally, a gauge should be of a size equal to the work low limit, or alternatively, high limit. However, the gauge maker must of course have a manufacturing tolerance, and this is generally equal to 10% of the work tolerance, and is disposed opposite to the direction of wear. This is illustrated in Fig 13.17 for a limit plug gauge.

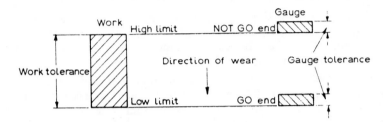

Fig 13.17   Tolerance for Plug Gauge.

Figure 13.17 shows that the direction of wear is such that both ends of the gauge will wear smaller. Hence, the gauge tolerance is disposed above the work low limit and high limit in order to extend the life of the gauge. This basic principle should always be adhered to. The British Standard Specification for gauge tolerances is BS 969 (plain limit gauges: limits and tolerances) which covers plain gauges only, and not screw thread or spline gauges for example. This standard recommends that the same gauge is used both for *inspection* and *workshop* purposes, i.e., in either case the gauge tolerance is the same. At one time different gauges were recommended for inspection use with the tolerances disposed rather differently. BS 1044: Part 1: 1964 (Plug, ring and caliper gauges) is a comprehensive standard which includes gauges referred to in this section.

Exercises 13

1. What is meant by quality? How much of it is determined by the designer and how much by the manufacturer?

2. 'A product cannot be designed both for appearance and reliability.' Discuss this statement.

3. What is a specification, and why is it so important to the production cycle?

4. What is meant by the term 'designing for production'? Discuss the implications of such a philosophy upon a manufacturing organization.

5. Define standardization, simplification, specialization and interchangeability, and show the relationship between each.

6. (a) What are preferred numbers?

(b) A product is to be marketed in a range of sizes from $\frac{1}{2}$ kW to 4 kW, and are to be based upon the R10 series.

Estimate the range of sizes, rounded off to the nearest 0·2 kW.
(Ans. Range has 10 sizes which are: 0·5, 0·6, 0·8, 1·0, 1·2, 1·6, 2·0, 2·6, 3.2 and 4·0 kW respectively)

7. Discuss the essential requirements for a system of interchangeability, and clearly show why such a system has economic advantages.

8. Define: (i) basic size, (ii) tolerance, (iii) limits, (iv) clearance fit, (v) interference fit, (vi) transition fit, (vii) fundamental deviation.

In terms of the above discuss and compare the Newall and BS 4500 systems of limits and fits.

9. (a) A hole and shaft have a basic size of 30 mm, and are to have a maximum clearance of 0·02 mm and a minimum clearance of 0·01 mm. The hole tolerance is to be 1·5 times the shaft tolerance. Calculate, with the aid of a diagram, the limits for both hole and shaft, using a hole basis system. (b) Calculate the same limits using a shaft basis system.

(Ans. (a) Shaft 29·99 mm   29·986 mm
      Hole 30 mm   30·006 mm
   (b) Shaft 30 mm   29·996 mm
      Hole 30·01 mm   30·016 mm)

10. Using your own component example, show an example of the effect of tolerance build-up.

11. What is meant by geometric tolerances, and how can they affect the functioning of a component? Discuss MMC and show how this has a bearing upon interchangeability.

12. (a) What is Taylor's principle of limit gauging?

(b) Design a gauge(s) to this principle for checking a square peg having limits of 25·00 mm and 24·97 mm. Sketch the gauge(s) and give all major dimensions if gauge tolerances are to be 10% of the work tolerance.

13. Draw and dimension a plug gauge to Taylor's principle suitable for checking a bore of 27·50 mm dia. × 45 mm long. Allowance should
                   27·52
be made for a gauge making tolerance.

Further Reading

1) 'Quality, its Creation and Control.' *The Institute of Production Engineers*, 1958.

2) Willsmore A. W. *Product Development and Design*. Pitman.

3) 'Design for Production—Productivity Report.' *British Productivity Council*.

4) Eilon S. *Elements of Production Planning and Control.* The Macmillan Co.

5) Lissaman A. J. and Martin S. J. *Principles of Engineering Production.* Hodder and Stoughton Ltd.

6) Gladman C.A. 'Design for Production.' *C.I.R.P. Annals,* 1968.

CHAPTER 14

Statistical Quality Control

14.1 QUALITY CONTROL

THIS CHAPTER is concerned with the control of quality upon the shop floor. Since the war there has been an increasing use of statistical techniques for this purpose, so much so, that the term *quality control* is often taken to mean the control of quality by statistical means. The theory of statistics and probability are brought together for this purpose, and the ideas behind the theory will be discussed in Sections 14.2 to 14.6. This is a very large subject indeed, and we can only introduce the principle, asking the reader to accept the theory used without proofs being given. Many fine text books are available on the subject, some of which are referred to in the 'Further Reading' section.

Statistical sampling techniques enable the inspector upon the shop floor to take controlled random samples of components which will then be checked by measurement or gauging. The quality of the whole batch of work produced will then be judged by the results of the sample, i.e., if the number of rejects found in the sample is too high then the quality of the whole batch is unacceptable; and vice versa. It is rather like taking a bite out of a large cake, and judging the quality of the whole cake upon the evidence obtained from tasting one mouthful, rather than eating the whole cake in order to be certain. This is where the idea of probability comes in. There is obviously a probability that the quality of the one mouthful of cake is different from the total quality level of the whole cake; either better or worse. The statistical part of the exercise is in collecting the data from the samples, in order that the total quality level can be assessed within a certain degree of probability. Hence, the inspector is as much concerned with numbers, charts, data, etc., as he is with measuring instruments and gauges.

The alternative to the above is to carry out 100% inspection such that every component leaving the production line is checked. This would appear to be more expensive, although the implication is that the results would have more certainty. However, this has been proved to be wrong, and there is evidence to show that 100% inspection does not guarantee that all unacceptable work will be found and rejected.

This is particularly so in the case of repetitive and boring activities, and let us face it, much engineering manufacturing work is in this category. In certain classes of work, where the part must be destroyed in order to check its quality, statistical sampling techniques must be used.

In addition to shop floor inspection, a manufacturing organisation can use sampling techniques for checking incoming goods, such as raw material, bought out parts and sub-assemblies for example. The two areas of application are shown at Fig 14.1.

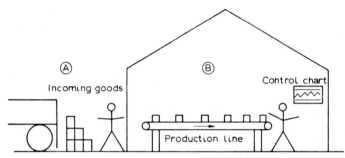

Fig 14.1 Areas of Application of Sampling Techniques.

From the sketch, 'A' is concerned with acceptance sampling, which is dealt with in Sections 14.10 to 14.12, and 'B' is concerned with control charts, which are dealt with in Sections 14.7 to 14.9.

14.2 MANUFACTURING VARIABILITY — DISTRIBUTION

As we saw in the last chapter, we must tolerate variabilities of size within certain limits during manufacture. No two parts can be produced with identical measurements. There will be variation in the measured sizes of parts due in part to the inherent inaccuracies of the manufacturing process, and also partly due to inaccuracies of the measuring equipment and its application.

Histograms

Let us assume that we measure the lengths of nine components in a sample, say, and obtain the following results:

1·8, 1·7, 1·8, 1·9, 2·0, 1·6, 1·8, 1·7, and 1·9 m respectively.

If we plot these results on a bar-diagram having the measured dimension (class) on the horizontal ordinate, and the frequency of occurrence of each dimension on the vertical ordinate, we shall construct a *histogram*. This is shown at Fig 14.2.

The histogram shown has the lengths of the parts grouped according to the frequency with which a length occurs. The difference between each value on the horizontal ordinate is called a class-interval (in this case

0·1 m), and each value is represented at the centre of the class interval as shown. This is a useful and simple way of representing measured data, and it can be seen at a glance, for example, that the arithmetic mean value of the dimensions is 1·8 m.

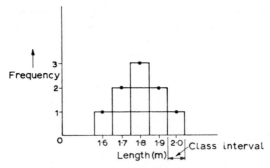

Fig 14.2 Histogram.

The degree of accuracy of the measurement of dimensions will affect the grouping and hence the histogram shape. If the lengths are measured now to a degree of accuracy of 0·05 m say, then every one of the nine results could be different and the histogram might appear as shown at Fig 14.3(a).

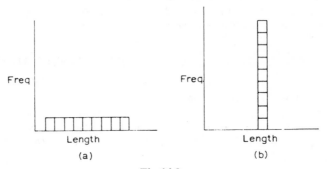

Fig 14.3

Again, if a very crude accuracy is now used, and the dimensions are grouped to the nearest metre, then the histogram would appear as Fig 14.3(b). The importance of this point will become self-evident in the next section.

Frequency Distribution

In statistical work it is more convenient to represent a group of data by means of a *frequency distribution* rather than a histogram. The ordinate scales are the same, but if we draw a smooth curve through the top centre points of the histogram bars (these points being shown by dots in Fig 14.2), we shall obtain a frequency distribution as shown at Fig 14.4.

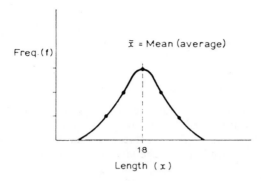

Fig 14.4 Frequency Distribution.

This frequency distribution is a graph having the same mean value as the histogram, i.e., 1·8 m. The measurement of an individual item is denoted by x, and $\bar{x}$ represents the arithmetic mean of a sample of n values. The frequency of occurrence of each x value is denoted by f.

This symmetrical bell-shaped type of distribution is typical of many engineering processes, the majority of the component measurements being grouped close to the mean, and fewer measurements being either smaller or larger than the mean. In practice, larger samples than nine would be required (say $n = 50$ to 100) in order to obtain a decisive shape for the frequency distribution.

14.3 MEAN AND SPREAD

Reasonably symmetrical distributions of the type shown at Fig 14.4 can be completely defined by (i) the *average* value, and (ii) the *variability* of the x values about this average value. Given these two parameters, which can be calculated from the collected data, it is unnecessary to plot all the individual results on a graph in order to obtain the shape of the distribution. Let us consider each.

i) Average
The *arithmetic mean* is most commonly used to give the mean value, $\bar{x}$ where:

$$\bar{x} = \frac{\Sigma x}{n}$$

Example 14.1
Estimate the mean of the x values, making up the distribution shown at Fig 14.4.

Solution

$$\text{mean} = \bar{x} = \frac{\Sigma x}{n}$$

$$= \frac{1 \cdot 6 + 1 \cdot 7 + 1 \cdot 7 + 1 \cdot 8 + 1 \cdot 8 + 1 \cdot 8 + 1 \cdot 9 + 1 \cdot 9 + 2 \cdot 0}{9}$$

$$= \frac{16 \cdot 2}{9} = 1 \cdot 8$$

ii) Variability

This refers to the spread or scatter of x values about the mean and is most commonly measured by the *range,* or the *standard deviation.*
Range (w). This is a simple but rather crude measure of spread, being equal to the difference between the largest and smallest x values, i.e.,

$$w = x_{max} - x_{min}$$

Example 14.2

Calculate the range of the distribution shown at Fig 14.4.

Solution

$$\text{Range} = w = x_{max} - x_{min}$$
$$= 2 \cdot 0 - 1 \cdot 6 = 0 \cdot 4$$

Standard Deviation(s)

This is a better measure of spread, because it is based upon all the sample results, not just the extreme values as is the case with the range. However, the arithmetic is more difficult, and it is often not practicable for inspectors to have to calculate it on the shop floor.

The standard deviation is a *root mean square* (R.M.S.) value and is defined:

$$s = \sqrt{\frac{\Sigma (x - \bar{x})^2}{n}}$$

Example 14.3

Calculate the standard deviation of the distribution shown at Fig 14.4.

Solution

This type of problem is best carried out in tabular form, each column being completed in turn:

x	$x - \bar{x}$	$(x - \bar{x})^2$
1·6	− 0·2	0·04
1·7	− 0·1	0·01
1·7	− 0·1	0·01
1·8	0	0
1·8	0	0
1·8	0	0
1·9	0·1	0·01
1·9	0·1	0·01
2·0	0·2	0·04

$$\Sigma (x - \bar{x})^2 = 0 \cdot 12$$

$$\therefore \quad \text{standard deviation} = s = \sqrt{\frac{\Sigma (x - \bar{x})^2}{n}}$$

$$= \sqrt{\frac{0.12}{9}} = \sqrt{0.0133} = 0.115$$

This method of calculating s working from first principles, is tedious when one is dealing with a large sample. Therefore, it is advisable to use a quick method of estimating $\bar{x}$ and s when n is large, and this can be done by making use of a 'dummy number', t. This is best illustrated by an example.

Example 14.4
A sample of 30 components were tested for strength (N), and the results obtained were as follows:

1 045	1 125	1 090	1 200	1 152
1 065	1 195	1 065	1 126	1 275
995	1 051	1 140	1 050	1 063
1 055	1 085	1 092	1 112	1 088
1 072	1 102	1 243	1 141	1 155
1 210	1 105	1 169	1 150	1 100

Estimate $\bar{x}$ and s for the sample.

Solution
The results must first be grouped according to frequency of occurrence of each x value, when a class interval has been decided upon. This can be done quite arbitrarily, and a little experience will soon lead to a choice of class interval giving a sensible result. The mid-point of each class is then taken as the x value, thus:

Class	x	f
900 – 999	949·5	1
1 000 – 1 099	1 049·5	12
1 100 – 1 199	1 149·5	13
1 200 – 1 299	1 249·5	4
		$n = 30$

This table represents the 30 strength values grouped according to their frequency, i.e., one value lies between 900 – 999, 12 values lie between 1 000 – 1 099, etc. 949·5 lies midway between 900 – 999, and so on. The class interval in this case has been arbitrarily chosen as 100.

With the data grouped in this manner $\bar{x}$ and s can be quickly estimated using the following formulae:

$$\bar{x} = x_0 + c \left(\frac{\Sigma ft}{\Sigma f} \right)$$

$$s = c \sqrt{\frac{\Sigma ft^2}{\Sigma f} - \left(\frac{\Sigma ft}{\Sigma f} \right)^2}$$

Where $c =$ class interval

$x_0 =$ estimated mean

$t =$ 'dummy number'

These are not nearly as bad as they look, and one has now only to choose any sensible x value which appears to be close to $\bar{x}$ in value, this chosen x value then being called x_0, (in this case, $x_0 = 1\,049 \cdot 5$).

Now set out a table, and complete the columns, thus:

x	f	t	ft	ft^2
949·5	1	−1	−1	1
1 049·5	12	0	0	0
1 149·5	13	1	13	13
1 249·5	4	2	8	16
$\Sigma f = 30$			$\Sigma ft = 20$	$\Sigma ft^2 = 30$

t must be made equal to zero, on the same line where x_0 has been chosen, (i.e., at $x = 1\,049 \cdot 5$). t is then made equal to −1, −2, −3, etc., and 1, 2, 3, etc., on each line, either side of the line in the table where $t=0$.

$$\text{Now } \bar{x} = x_0 + c \left(\frac{\Sigma ft}{\Sigma f} \right)$$

$$= 1\,049 \cdot 5 + 100(\tfrac{20}{30}) = 1\,116 \text{ N}$$

$$s = c \sqrt{\frac{\Sigma ft^2}{\Sigma f} - \left(\frac{\Sigma ft}{\Sigma f} \right)^2}$$

$$= 100 \times \sqrt{\frac{30}{30} - \left(\frac{20}{30} \right)^2} = 75 \text{ N}$$

If the reader is not yet convinced that this approximate method is simpler and quicker than working from first principles, then let him attempt a comparison and check the accuracy of the above answers!

Since the widespread adoption of pocket electronic calculators the advantages of 'quick' methods of statistical calculation are not now so attractive. However, the method shown above still has some educational value.

14.4 PROBABILITY

Probability is measured on a scale having limits of 0 and 1, where 0 represents absolute impossibility, and 1 represents absolute certainty. The symbols '*p*' and '*q*' are used to represent probability, thus:

$p =$ probability of an event happening.

$q =$ probability of an event *not* happening.

∴ If the probability of an event happening is absolute certainty, then

$p = 1$

A little thought will show that $p + q = 1$

If an unbiased coin is tossed, then it is absolutely certain that a head or tail will turn up, i.e., p (tossing head or tail) $= 1$

$$q \text{ (tossing head or tail)} = 0$$

$$\therefore p + q = 1 + 0 = 1$$

Common sense will show that the coin, if tossed, must turn up a head or a tail with equal probability,

$$\therefore p \text{ (tossing head)} = \tfrac{1}{2}$$

$$p \text{ (tossing tail)} = \tfrac{1}{2}$$

and the sum of the probabilities of all the possible occurrences (in this case two occurrences only) is 1.

We can now say that p can be defined thus:

$$p = \frac{\text{Number of outcomes favourable to the event}}{\text{Total number of possible outcomes}}$$

Apply this formulae to the coin tossing event:

$$p \text{ (tossing head)} = \frac{\text{Number of favourable outcomes (1)}}{\text{Total number of outcomes (2)}} = \tfrac{1}{2}$$

This does not mean that we would expect a head to be turned up 5 times for every 10 times the coin is tossed, say. In the long run, however, if the coin was tossed 1 000 000 times, we would expect *nearly* half of the tosses to show a head; the order in which the heads and tails showed to be at random, having no pattern.

Consider an example in which a sample *(n)* is taken from a batch *(N)*.

Example 14.5

A box contains four white counters and two black counters. What is the probability of (a) drawing out one white counter from the box, and (b) drawing out one black counter from the box, in separate, random draws.

Here $n = 1$ and $N = 6$.

Solution

a) Taking out one counter at random,

$$p \text{ (drawing white)} = \frac{\text{Number of outcomes favourable to drawing white}}{\text{Total number of possible outcomes}}$$

$$= \frac{4}{6} = \frac{2}{3}$$

b) The answer here will depend upon whether the first counter withdrawn at (a) is replaced or not. If not:

$$p \text{ (drawing black)} = \tfrac{2}{5}$$

If it is:

$$p \text{ (drawing black)} = \tfrac{2}{6} = \tfrac{1}{3}.$$

The question of whether sampled parts are replaced or not is important. In practice, as parts are drawn out for sampling from a batch (or lot), they will not be replaced until sampling is complete. Every contingency is covered by using the following formulae, which are followed by an example.

i) Each item replaced after drawing (order of drawing is taken into account),

$$\text{Number of outcomes} = N^n$$

ii) Each item withdrawn and not replaced (order of drawing is taken into account),

$$\text{Number of outcomes} = {}^N P_n = \frac{N!}{(N-n)!} \text{ (This is a } permutation)$$

iii) Each item withdrawn and not replaced (no account taken of order of drawing),

$$\text{Number of outcomes} = {}^N C_n = \frac{N!}{(N-n)!\,n!} \text{ (This is a } combination)$$

Example 14.6

A box contains three counters numbered 1, 2 and 3 respectively. Two counters are withdrawn, separately, and at random.

Calculate the total number of possible ways in which the two counters out of three may be withdrawn, if

a) Each counter is replaced after drawing, and order of drawing is taken into account.

b) Each counter is *not* replaced after drawing, and order of drawing is taken into account.

c) Each counter is *not* replaced after drawing, and no account is taken of order of drawing.

Solution

$$\text{Here } n = 2 \text{ and } N = 3$$

a) total number of ways $= N^n = 3^2 = 9$

$$\begin{aligned}
Check : \ &(1,1,) \ (1,2,) \ (1,3,) \\
&(2,1,) \ (2,2,) \ (2,3,) = 9 \text{ ways.} \\
&(3,1,) \ (3,2,) \ (3,3,)
\end{aligned}$$

As order of drawing is taken into account, 1,2, is counted as a different outcome than 2,1.

b) total number of ways $= {}^N P_n = \dfrac{N!}{(N-n)!}$

$$= \frac{3 \times 2 \times 1}{1} = 6$$

$$\begin{aligned}
Check : \ &(1,2,) \ (1,3,) \\
&(2,1,) \ (2,3,) = 6 \text{ ways.} \\
&(3,1,) \ (3,2,)
\end{aligned}$$

If 1 is withdrawn first and not replaced, then it can only be followed by a 2 or 3, and likewise if 2 or 3 is withdrawn first. This means there are three ways less.

c) This is the practical shop floor sampling case, where items are not usually replaced immediately after sampling, and order of drawing does not usually matter.

$$\text{total number of ways} = {}^{N}C_{n} = \frac{N!}{(N-n)!\,n!}$$

$$= \frac{3 \times 2 \times 1}{1 \times 2 \times 1} = \frac{6}{2} = 3$$

Check: (1,2,)
       (1,3,) = 3 ways
       (2,3,)

There are two laws of probability we should consider before leaving this section, these being the addition law and the multiplication law.

Addition Law

If events are mutually exclusive (i.e., if one event happens the other cannot) then the total probability (P) of any one of the events happening is the sum of the separate probabilities, i.e.,

$$P = p_1 + p_2 + p_3, \text{ etc.}$$

For example, the probability of throwing 5 or 6 with one throw of a dice $= \frac{1}{6} + \frac{1}{6}$

$$\therefore P = \tfrac{1}{3}.$$

We add the separate probabilities of throwing either 5 or 6 separately to give the total probability, because if the 5 is turned up, then the 6 cannot be turned up. The separate events of a 5 or 6 turning up are mutually exclusive.

Multiplication Law

If events are *not* mutually exclusive then the total probability is the product of the separate probabilities, i.e.,

$$P = p_1 \times p_2 \times p_3, \text{ etc.}$$

For example, the probability of turning up a head each time for three separate tosses of a coin $= \frac{1}{2} \times \frac{1}{2} \times \frac{1}{2}$

$$\therefore P = \tfrac{1}{8}$$

The events are not mutually exclusive, because if a head is turned up at the first toss, a head can still be turned up again at the second toss, and so on. Therefore, the separate probabilities are multiplied.

(Note that the answer would be the same if the tosses were carried out simultaneously using three separate coins, rather than successively using one coin.)

An example is now given of the application of these laws.

Example 14.7

Two components are chosen at random from a batch of ten components, without replacement. The batch contains three defectives.
Calculate:

a) p (both components are defective)
b) p (both components are *not* defective)
c) p (one component is defective, and one is not).

Solution

a) $p = \frac{3}{10} \times \frac{2}{9} = \frac{6}{90} = \frac{1}{15}$.

b) $p = \frac{7}{10} \times \frac{6}{9} = \frac{42}{90} = \frac{7}{15}$.

c) There are two methods of calculating p (one of each);
Firstly, as this is the only other possibility apart from (a) and (b), then:

$$p = 1 - (\tfrac{1}{15} + \tfrac{7}{15}) = \tfrac{7}{15}.$$

Alternatively, applying the multiplication and addition laws from first principles:

$$p = (\tfrac{3}{10} \times \tfrac{7}{9}) + (\tfrac{7}{10} \times \tfrac{3}{9}) = \tfrac{21}{90} + \tfrac{21}{90} = \tfrac{7}{15}.$$

14.5 RELATIVE FREQUENCIES

We will endeavour to show now how the statistical data (in the form of frequency distributions) and probability principles are joined in the field of quality control.

The frequency distribution we have considered so far (Fig 14.4) is a continuous distribution because the x values can be of any magnitude upon the horizontal, continuous scale. Another type of distribution concerned with sampling is a discrete distribution, in which the x values can only be positive integers, i.e., positive whole numbers.

As an example, consider the distribution shown at Fig 14.5.

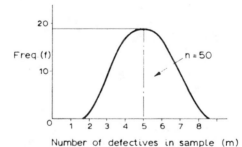

Fig 14.5   Frequency Distribution.

Along the horizontal scale is plotted the number of defectives found in a sample (from two to eight), the distribution showing the results of many samples. This is a discrete distribution in which the sampling is done by gauging, the components in the sample being judged defective or acceptable. As one cannot have other than whole numbers of defectives, the *x* values are positive integers. Again, for most engineering processes, this will be a reasonably symmetrical, bell shaped distribution.

Now consider the distribution plotted with a vertical ordinate having a *relative frequency* scale. The distribution will look the same, but the results will have to be expressed differently. Say the distribution shown at Fig 14.5 contains 50 results (i.e., 50 frequencies of occurrence of every *x* value) and that five defectives for example occurred during sampling with a frequency of 19. Then on a relative frequency scale, 5 on the horizontal scale will be opposite the value $\frac{19}{50}$ upon the vertical scale. This is shown at Fig 14.6.

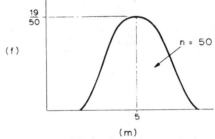

Fig 14.6   Relative Frequency Distribution.

If every *x* value is expressed at a relative frequency then the total of the relative frequencies for the distribution must equal $\frac{50}{50} = 1$.

Now to get to the point—if the sampling data forming the distribution has been taken from a stable process over a long period of time, then the relative frequency distribution effectively becomes a *probability distribution,* in the long run. In our sample, it could be stated that

$$p \text{ (getting five defects in sample of size } n) = \frac{19}{50}$$

or again:

$$p \text{ (getting two to eight defects in sample of size } n) = \frac{50}{50} = 1$$

i.e., absolute certainty.

There are two important, standard discrete distributions which can be expressed mathematically and used as probability distributions, if sufficient data has been collected to show that the process has the inherent characteristics to allow this. The distributions are the *binomial,* and the *poisson.* In one chapter we can only consider each very briefly with the aid of one example.

Binomial Distribution

This is defined as $(p + q)^n$

$$= q^n + (^nC_1)q^{n-1}p + (^nC_2)q^{n-2}p^2 + (^nC_3)q^{n-3}p^3 \ldots p^n$$

This distribution would be applicable for sampling from a continuous process where all individual articles selected from sampling have the same probability of being defective. [This implies an infinitely large batch N, and p (one item being defective) being unaffected by p (another item being defective). This is clearly not so with small batches, where parts are being withdrawn and not replaced.]

Example 14.8
The probability that a single component in a continuous process is defective is 0·4. A sample of six is drawn at random from a large batch of the components. Calculate the probability of finding the following number of defective components (m) in the sample: (a) 0, (b) 1, (c) 2, (d) 3, (e) 4, (f) 5, and (g) 6.

Solution
$$\text{Here } n = 6, p = 0·4, q = 0·6$$

Working from first principles would make this a difficult problem, but as the binomial process is appropriate, it can be used to estimate the probabilities. Each term of the expression gives a probability, thus:

$$\underbrace{\left[(0·6)^6\right]}_{\text{(a)}} + \underbrace{\left[\frac{6\times5\times4\times3\times2\times1}{5\times4\times3\times2\times1}\times(0·6)^5\times(0·4)\right]}_{\text{(b)}} +$$

$$\underbrace{\left[15\times(0·6)^4\times(0·4)^2\right]}_{\text{(c)}} + \underbrace{\left[20\times(0·6)^3\times(0·4)^3\right]}_{\text{(d)}} + \underbrace{\left[15\times(0·6)^2\times(0·4)^4\right]}_{\text{(e)}} +$$

$$\underbrace{\left[6\times(0·6)\times(0·4)^5\right]}_{\text{(f)}} + \underbrace{\left[(0·4)^6\right]}_{\text{(g)}}$$

∴ The separate probabilities are:

$$0·047 \;+\; 0·187 \;+\; 0·311 \;+\; 0·277 \;+\; 0·138 \;+\; 0·037 \;+\; 0·004$$
$$p\,(0\text{ def.})\; p\,(1\text{ def.})\; p\,(2\text{ def.})\; p\,(3\text{ def.})\, p\,(4\text{ def.})\; p\,(5\text{ def.})\; p\,(6\text{ def.})$$

If these values are plotted as a distribution with m values on the horizontal ordinate, and p values on the vertical ordinate, the mean value $= np = 6 \times 0·4 = 2·4$. It will be appreciated that all the separate p values total 1 (allowing for rounding off).

Poisson Distribution
This is defined as:

$$e^{-z} + ze^{-z} + \frac{z^2e^{-z}}{2!} + \frac{z^3e^{-z}}{3!}, \text{ etc.}$$

each of the terms of the distribution giving the probability of occurrence of 0, 1, 2, 3, etc. events, where $z =$ the average number of occurrences.

This distribution is similar to the binomial but is used where the value of n is not known. For example, this case would occur during the inspection of finish of panels where different types of flaws found could reject the panel.

Example 14.9

Panels being inspected for finish display on the average two flaws. Calculate the probability of (a) 0, (b) 1, (c) 2, and (d) 3 flaws being found.

Solution

$$z = 2$$

Probabilities are given by:

$$e^{-z}\left(1 + z + \frac{z^2}{2!} + \frac{z^3}{3!}\right)$$

$$= e^{-2}\left(1 + 2 + \frac{4}{2} + \frac{8}{6}\right)$$

$$= 0.135\left(1 + 2 + 2 + \frac{4}{3}\right)$$

∴ the separate probabilities are:

0·135	+	0·270	+	0·270	+	0·180
p (0 flaws)		p (1 flaw)		p (2 flaws)		p (3 flaws)

It can be seen that p (four flaws or more) is small being $1 - 0.855 = 0.145$.

14.6 NORMAL DISTRIBUTION

A third probability distribution which is important to engineers is the *normal* (or Gaussian) *distribution*. It is a continuous distribution having a symmetrical bell shape, like a distribution of dimensions obtained from a sample of mass produced articles. The normal distribution will therefore enable us to forecast the probability of recurrence of certain events as a result of sampling, when the sample results are measured from an engineering process.

The normal distribution has an area of 1 under the curve, the mean being equal to 0, and the standard deviation equal to 1. The values of the area under the curve are shown at Fig 14.7 for various values of the standard deviation.

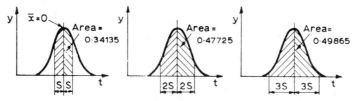

Fig 14.7 Normal Distribution.

At Fig 14.7(a) it can be seen that the area under the curve subtended by the mean and one standard deviation equals 0·341 35, or the area subtended by two standard deviations = 68·27%. Similarly, the area subtended by four standard deviations = 95·45%, and by six standard deviations = 99·73%. Therefore we can say that p (any value falling between limits of $\pm s$) = 0·682 7, and so on for other limits. Let us show by means of two examples how this probability distribution can be applied to measured results.

Example 14.10

A distribution of sample results from a mass production process is obtained, this distribution being symmetrical, having $\bar{x} = 10$ mm and $s = 1·3$ mm. What is the probability of a component size of between 10 and 12·6 mm being obtained in a random sample?

Solution

Find a value of t for the normal distribution which being similar to the process distribution can be used to obtain p.

$$t = \frac{x - \bar{x}}{s} = \frac{12·6 - 10}{1·3} = \frac{2·6}{1·3}$$
$$= 2$$

This means that 12·6 is two standard deviations away from the mean, as shown in Fig 14.8.

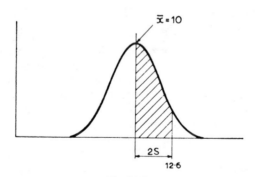

Fig 14.8

We know that the area under the normal curve between limits of the mean and two standard deviations is 47·725% (see Fig 14.7), therefore p (size between 10 and 12·6) = 0·477 approx.

Example 14.11

In Example 14.10 calculate
a) p (size between 6·1 and 12·6), and
b) p (any size < 6·1 and > 12·6).

Solution

a) For value $6 \cdot 1$ mm $t = \dfrac{x - \bar{x}}{s} = \dfrac{6 \cdot 1 - 10}{1 \cdot 3}$

$$= \frac{-3 \cdot 9}{1 \cdot 3} = -3$$

(Note. Values to the left of the normal distribution mean are negative. This does not affect calculations).

$\therefore$ $6 \cdot 1$ is three standard deviations away from the mean, and the area under the curve bounded by the mean and three standard deviations is $0 \cdot 498\ 65$.

$\therefore$ p (size between $6 \cdot 1$ and $12 \cdot 6$)

$= 0 \cdot 498\ 65 + 0 \cdot 477\ 25 = 0 \cdot 975\ 9$.

b) To find the probability of any size occurring in a sample between $6 \cdot 1$ mm and $12 \cdot 6$ mm we should remember that the total area under the curve $= 1$ (i.e., absolute certainty)

$\therefore$ p (size $< 6 \cdot 1$ and $> 12 \cdot 6$)

$= (0 \cdot 500\ 00 - 0 \cdot 498\ 65) + (0 \cdot 500\ 00 - 0 \cdot 477\ 25)$

$= 0 \cdot 001\ 35 + 0 \cdot 022\ 75 = 0 \cdot 024\ 1$

(Alternatively, $p = 1 - 0 \cdot 975\ 9 = 0 \cdot 024\ 1$).

Any problems in which t comes out to other than a whole number, can be solved by using statistical tables such as Cambridge Elementary Statistical Tables, in which the areas under the normal curve for various values, are tabulated under the heading of *Normal cumulative probability function*.

Distribution of Sample Means

If samples of n parts are taken at random from a controlled process which gives approximately a normal distribution, the sample means will themselves be normally distributed having a smaller amount of spread or scatter. Note that any distribution of means ($\bar{x}$) will tend to be more normal in shape than the original distribution of individual (x) values.

The average value of a distribution of means is $\bar{X}$ (grand mean), and the standard deviation is $s_n = \dfrac{s}{\sqrt{n}}$

Example 14.12

Samples of 30 components give a distribution of individual x values having a standard deviation of 75 (see Example 14.4). If a distribution is compiled of the mean values of several samples of 30, estimate the standard deviation of the distribution of means.

Solution

Standard deviation of distribution of means $= s_n = \dfrac{s}{\sqrt{n}}$

$$= \frac{75}{\sqrt{30}} = 13 \cdot 7.$$

It can be seen from this that the distribution of means lies closer to the average having less spread, and as n increases so s_n decreases. A little thought will show that this is only to be expected. The principle of distribution of means is important to an understanding of control charts.

Distribution of Sample Ranges

Similarly, the standard deviations or ranges of distributions of individual sizes, could themselves be distributed. The distribution of ranges is important, because it is the one used for control charts, w being easier to calculate than s. This distribution will be found to be skew, i.e., the distribution of results is not symmetrically disposed around the mean $(\bar{w})$, the spread on one side of the mean being greater than the spread on the other side.

14.7 CONTROL CHARTS FOR ATTRIBUTES (P or m CHARTS)

This type of control chart is used for manufactured articles which can be classed as *good* or *bad*, e.g., machined metal parts which are being inspected with limit gauges and being classified as acceptable or defective. The work is random sampled regularly, the results being plotted upon the control chart. The main purpose of the chart is to indicate changes in quality so that adjustments can quickly be made to correct the process before large quantities of scrap are produced, i.e., in effect the chart gives advance warning of the commencement of a trend toward the production of an increasing number of defective articles.

There are two alternative types of chart available for the *control of attributes*, these being Fraction defective (P) chart, and Number defective (m) chart respectively. Each will be considered in turn.

Fraction defective chart

The compilation of this chart is perhaps best considered in two steps:

Step 1

Collect data from the process by taking samples of size n and counting the number of defectives in each sample, then calculate the *fraction defective* which $= P = \dfrac{\text{number of defects in sample}}{n}$

Draw up a frequency distribution of P values which should be approximately normal having a mean value of

$$\bar{P} = \frac{\text{total number of defects found}}{\text{total number of parts sampled } (\Sigma n)}$$

Example 14.13(a)

Twenty samples of parts were taken from a production line for gauging, each sample containing 100 parts. The following number of defects were found in each sample:

$$3, 4, 5, 5, 5, 5, 4, 7, 6, 7,$$
$$6, 6, 4, 3, 5, 8, 5, 4, 6, 5.$$

Calculate $\bar{P}$ and draw the distribution.

Solution

$$\bar{P} = \frac{\text{total number of defects}}{\text{total number sampled}} = \frac{103}{2\,000} = 0.052.$$

The distribution is shown at Fig 14.9.

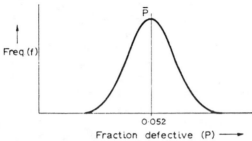

Fig 14.9 P — Distribution.

Step 2

Turn the P distribution through 90°, and draw the *inner control limits* (warning), and the *outer control limits* (action) on the control chart. These limits are positioned at two standard deviations (2s), and three standard deviations (3s) respectively from the mean ($\bar{P}$). The standard deviation for a P distribution is calculated thus:

$$s = \sqrt{\frac{\bar{P}(1 - \bar{P})}{n}}$$

The resulting chart is called a P chart.

Example 14.13(b)

Calculate the limits, and draw the control chart for the process in Example 14.13(a).

Solution

$$s = \sqrt{\frac{\bar{P}(1 - \bar{P})}{n}} = \sqrt{\frac{0.052 \times 0.948}{100}}$$

$$= \sqrt{0.000\,5} = 0.022.$$

$$\therefore 2s = 0.044 \text{ and } 3s = 0.066.$$

The control chart is shown at Fig 14.10. (In practice, the distribution curve is not drawn upon the chart, only the limits being shown.)

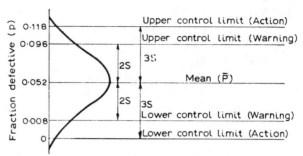

Fig 14.10 P Chart.

Regular, random samples are taken by the inspector, and checked, the *P* value for each sample being calculated and plotted upon the chart. We would expect (from our knowledge of the normal probability distribution) 99·73 % of the points plotted to fall within the outer limits, and 95·45 % to fall within the inner limits, due to the natural variability of the process. To put it another way, two points in 1 000 should fall outside the outer limits, and 45 points in 1 000 should fall outside the inner limits, by the laws of probability. If this is so, the process is deemed to be in *statistical control*. If more points start to fall outside the limits than can be attributed to chance, then the warning should be taken followed by action (if necessary) in order to correct the process. This means that an assignable cause for the drift outside the limits must be found, such as tool wear, for example. Points falling below the lower limits indicate an improvement in quality because the trend is then towards less rejects being found, so the lower limits are not as important as the upper.

Example 14.13(c)
After the control chart in Fig 14.10 is compiled, the following defects from the process for 10 samples of 100 are obtained. Plot the results on the chart.

3, 4, 4, 3, 4, 5, 6, 10, 11, 11, defects respectively.

Solution
Convert the results of sampling into *P* values, where $P = \frac{3}{100}, \frac{4}{100}, \frac{4}{100}$, etc. The results are then:
0·03, 0·04, 0·04, 0·03, 0·04, 0·05, 0·06, 0·10, 0·11, and 0·11.
These are shown plotted on the chart at Fig 14.11.

The process shows a drift towards running out of statistical control which if continued would require action, thus saving an accumulation of scrap. If the original sample results are plotted, then these will be seen to be in statistical control with the normal amount of variation inherent in the process. Control charts exclude guesswork, leading to positive action being taken only when required.

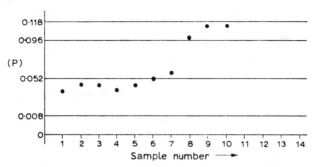

Fig 14.11 P Chart showing plotted results.

Number defective chart.

The procedure for this alternative type of chart is similar to that described for a '*P*' chart, except that the limits can be read directly from a table.

Step 1

Collect data from the process by taking samples at random, and counting the number of defectives in each sample, then calculate the mean number of defects per sample $(\overline{m})$, where:

$$\overline{m} = \frac{\text{total number of defects found}}{\text{total number of samples}}$$

Step 2

Determine the upper control limits from the table at Fig 14.12. The resulting chart is called an '*m*' chart. As stated earlier, lower limits are of no great importance in the context of control charts for attributes. It will be seen that the table at Fig 14.12 only covers *m* values up to a maximum to 2·0. For values higher than this, one needs to use a Poisson distribution cumulative probability chart.

Average number of defects expected in the sample $(\overline{m})$	U.C.L. (*Warning*)	U.C.L. (*Action*)
0·6	2·5	3·8
0·8	2·9	4·4
1·0	3·3	4·8
1·2	3·7	5·2
1·4	4·0	5·6
1·6	4·4	6·1
1·8	4·7	6·5
2·0	5·0	6·8

Fig 14.12 Limits for number defective control charts.

The following simple example shows how an '*m*' chart is constructed.

Example 14.14

a) The results of 19 random samples for number of defective items is as follows:

$$0, 2, 2, 3, 2, 1, 1, 3, 2, 2, 4, 3, 3, 2, 4, 1, 0, 1, 2.$$

Determine the limits.

b) The results of a further 25 samples of number defective taken subsequently are as follows:

$$2, 2, 2, 0, 1, 3, 1, 4, 2, 2, 3, 3, 0, 1, 2, 2, 3, 2, 0, 2, 6, 1, 5, 1, 4.$$

Plot these results upon the '*m*' chart.

Solution

a)
$$\bar{m} = \frac{\text{total number of defects}}{\text{total number of samples}}$$

$$= \frac{38}{19} = 2\cdot0$$

From table at Fig 14.12

$$\text{(Action) U.C.L.} = 6\cdot8$$
$$\text{(Warning) U.C.L.} = 5\cdot0$$

b) The control chart is shown at Fig 14.13 with the plotted results.

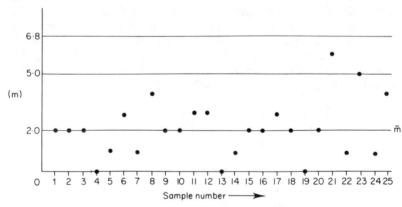

Fig 14.13   '*m*' Chart showing plotted results.

14.8 CONTROL CHARTS FOR VARIABLES ($\bar{x}$ and w CHARTS)

These types of control charts are used for a manufactured part which the inspector checks by measurement, not gauging. This is less common, and. of course this method of control is more expensive. However, it reveals much more about the behaviour of the process. The justification for using these types of control charts is exactly as stated in the last section.

With sample data in the form of measurements, it is possible to use a control chart showing deviation in the process mean ($\bar{x}$ *chart*), and one showing deviation in variability (*w chart*). The procedure for compiling these charts is similar to that for P charts, in that data has to be collected initially to establish the value of $\bar{X}$ and $\bar{w}$ from a series of samples (usually five or 10 in number). Control limits can then be calculated, and the charts will appear as shown at Fig 14.12.

Points plotted on the charts will indicate any tendency for the process to go out of statistical control. Any change of central tendency shown on the $\bar{x}$ chart may be due to tool wear, temperature increase, new work materials, etc. Any change of variability shown on the *w* chart is more difficult to account for, but may be due to wear in machine bearings or slides, affect of careless operator (where one is employed), etc. The L.C.L.'s on this chart (like the P chart), are less important than the U.C.L.'s, because a trend towards the lower limits indicates an improvement of quality as the variability is less.

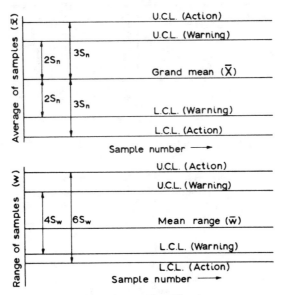

Fig 14.14 $\bar{x}$ and *w* Charts.

With $\bar{x}$ and *w* charts, the position of the limits is not calculated from first principles because the computation of the standard deviations can be very tedious as was seen in earlier sections. Tabulated data shown at Fig 14.15 and 14.16 is used (ref. BS 2564), this making the calculation of the standard deviation of the distribution of means (s_n), and the calculation of the standard deviation of the distribution of ranges (s_w) unnecessary.

FACTORS USED IN $\bar{x}$ CHARTS

Sample size n	Warning factor $A^1_{0.025}$	Action factor $A^1_{0.001}$
2	1·229	1·937
3	0·668	1·054
4	0·476	0·750
5	0·377	0·594
6	0·316	0·498
7	0·274	0·432
8	0·244	0·384
9	0·220	0·347
10	0·202	0·317

Fig 14.15

FACTORS USED IN w CHARTS

Sample size n	Upper action factor $D^1_{0.999}$	Upper warning factor $D^1_{0.975}$	Lower warning factor $D^1_{0.025}$	Lower action factor $D^1_{0.001}$
2	4·12	2·81	0·04	0·00
3	2·98	2·17	0·18	0·04
4	2·57	1·93	0·29	0·10
5	2·34	1·81	0·37	0·16
6	2·21	1·72	0·42	0·21
7	2·11	1·66	0·46	0·26
8	2·04	1·62	0·50	0·29
9	1·99	1·58	0·52	0·32
10	1·93	1·56	0·54	0·35

Fig 14.16

Using this data, the control limits can be calculated thus:

$\bar{x}$ *Chart.*

$$\text{(Action) U.C.L.} = \overline{X} + A^1_{0.001}\overline{w}$$
$$\text{(Action) L.C.L.} = \overline{X} - A^1_{0.001}\overline{w}$$
$$\text{(Warning) U.C.L.} = \overline{X} + A^1_{0.025}\overline{w}$$
$$\text{(Warning) L.C.L.} = \overline{X} - A^1_{0.025}\overline{w}$$

w Chart.

$$\text{(Action) U.C.L.} = D^1_{0.999}\overline{w}$$
$$\text{(Action) L.C.L.} = D^1_{0.001}\overline{w}$$
$$\text{(Warning) U.C.L.} = D^1_{0.975}\overline{w}$$
$$\text{(Warning) L.C.L.} = D^1_{0.025}\overline{w}$$

Let us take one example in order to illustrate the procedure.

Example 14.15

Samples of five were taken at regular intervals from a process, 10 samples in all being taken. The results were as follows:

Sample No.	Measurements per sample (hundredths of one mm)				
1	747	748	747	749	748
2	748	749	750	748	749
3	749	748	750	748	749
4	749	749	750	750	751
5	749	749	750	750	751
6	749	750	751	749	750
7	750	750	751	751	750
8	751	750	750	750	752
9	751	751	752	751	751
10	751	752	752	753	751

Calculate warning and action limits for $\bar{x}$ and w Charts. Show the results of these samples plotted upon the charts.

Solution
The grand mean $\bar{X}$ of all 10 samples is found to be 7·499mm, and the mean range $\bar{w}$ is found to be 0·018 mm.

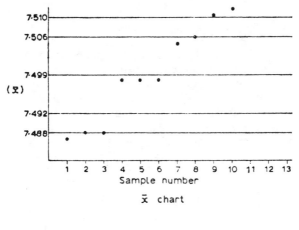

$\bar{x}$ chart

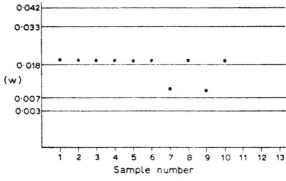

w chart

Fig 14.17 $\bar{x}$ and w Charts showing plotted results.

x̄ Chart.

(Action) U.C.L. $= 7\cdot499 + (0\cdot594 \times 0\cdot018) = 7\cdot510$ mm
(Action) L.C.L. $= 7\cdot499 - (0\cdot594 \times 0\cdot018) = 7\cdot488$ mm
(Warning) U.C.L. $= 7\cdot499 + (0\cdot377 \times 0\cdot018) = 7\cdot506$ mm
(Warning) L.C.L. $= 7\cdot499 - (0\cdot377 \times 0\cdot018) = 7\cdot492$ mm

w Chart.

(Action) U.C.L. $= 2\cdot34 \times 0\cdot018 = 0\cdot042$ mm
(Action) L.C.L. $= 0\cdot16 \times 0\cdot018 = 0\cdot003$ mm
(Warning) U.C.L. $= 1\cdot81 \times 0\cdot018 = 0\cdot033$ mm
(Warning) L.C.L. $= 0\cdot37 \times 0\cdot018 = 0\cdot007$ mm

The charts and plotted results are shown at Fig 14.17.

These charts show that the mean size was out of statistical control when the samples were taken, fluctuating from the low side of the mean to the high side. At the same time the variability or spread of the sampled dimensions is very constant.

14.9 RELATIVE PRECISION INDEX

We have said nothing yet about the relationship between the statistical control limits, and the tolerance called for in the specification. This is shown at Fig 14.18 where two examples are given:

a) A component having a wide dimensional tolerance being manufactured by a process having little variability.

b) A component having a narrow dimensional tolerance being manufactured by a process having great variability.

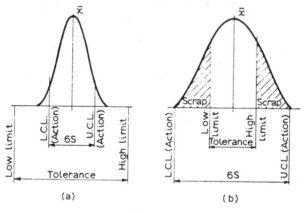

Fig 14.18

In case (a), the process is in effect too good for the specification. The statistical action limits have little meaning. It is possible to greatly vary x̄ and w without transgressing the specification limits.

In case (b), the process is too crude for the specification, much scrap being produced. Again, there is a wide divergence between the specification limits and the statistical limits.

This illustrates another important function of control charts, viz., they clearly show whether or not the process capability is compatible with the specification. The relationship between the two can be evaluated using the *relative precision index* (R.P.I.) which can be defined as the ratio of specification tolerance to average sample range, i.e.,

$$\text{R.P.I.} = \frac{\text{total specification tolerance}}{\text{average range}} = \frac{t}{\bar{w}}$$

The values of R.P.I. are described as low, medium or high, depending upon the size of the sample, n. Thus when $n = 4$, for example, the R.P.I. is low if < 3; medium if > 3 or < 4; high if > 4.

These values of R.P.I. can be interpreted thus:

Low R.P.I. — Unsatisfactory. Rejections are inevitable.

[Fig 14.18(b)]

Medium R.P.I. — Satisfactory.

High RPI — Parts are being produced with little variation.

[Fig 14.18(a)]

The classification of R.P.I. for sample sizes up to 6 is given in the table shown below.

Sample size	Low R.P.I.	Medium R.P.I.	High R.P.I.
(n)			
2	Less than 6	6 to 7	Greater than 7
3	Less than 4	4 to 5	Greater than 5
4	Less than 3	3 to 4	Greater than 4
5 and 6	Less than 2·5	2·5 to 3·5	Greater than 3·5

Example 14.16

The part produced in Example 14.16 has an average range of 0·018 mm. It also has a tolerance of 0·03 mm. Calculate the R.P.I. ($n = 5$).

Solution

$$\text{R.P.I.} = \frac{\text{total specification tolerance}}{\text{average range}}$$

$$= \frac{0 \cdot 03}{0 \cdot 018} = 1 \cdot 7$$

Therefore the R.P.I. value is low, being less than 2·5. This means that the process is unsatisfactory, and many rejects will be found in the samples.

14.10 ACCEPTANCE SAMPLING — SINGLE SAMPLING SCHEMES

As a form of protection, products passing into a factory must be assessed for quality, and as we have indicated, this can be done by sampling. Work is best delivered in batches of the same size from which controlled random samples can be taken. There is always a probability that an unsatisfactory batch will be passed, and that a good batch will be rejected. The former is called the *consumer's risk*, and the latter is called the *producer's risk*.

Acceptance sampling is usually by *attributes,* the parts being classed as acceptable or defective. The basis of all schemes is that: (i) *Samples* of size n are drawn from a batch or *lot* of size N, and inspected. (ii) The remainder $(N - n)$ of any lots which are rejected because of the results of the sample, are 100% inspected. The cost of such additional inspection to be borne by the consumer, or the producer by agreement.

A single sampling scheme is shown diagrammatically at Fig 14.17.

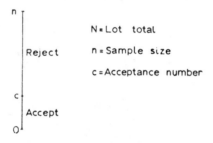

Single sampling scheme

Fig 14.19   Single sampling Scheme.

This can be interpreted as:

Inspect a sample n from lot N.

If O to C defects are found in the sample, accept the lot.

If more than C defects are found in the sample, reject the lot.

Operating Characteristic Curve

The characteristics of a sampling scheme are given by an *operating characteristic curve* (O.C.C.), an example of which is shown at Fig 14.20.

Referring to Fig 14.18 the symbols shown are:

α = producer's risk (usually 0·05 or 5%)

β = consumer's risk (usually 0·10 or 10%)

AQL = acceptable quality level (%)

$LTPD$ = lot tolerance percent defective (%)

The fraction defective (P) values along the horizontal ordinate are usually expressed as a percentage.

If α and β were of the values shown above, [5% and 10% respectively

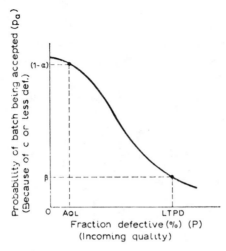

Fig 14.20 Operating Characteristic Curve.

giving $(1 - \alpha)$ a value of 95%], and $AQL = 3\%$ and $LTPD = 9\%$, say, the curve would then be interpreted thus:

The producer's risk is 5%, that batches containing 3% defectives (this quality level being acceptable to everyone) will be rejected.

The consumer's risk is 10% that batches containing 9% defectives (above the acceptable number of defectives) will be accepted.

An O.C.C. can be plotted, by finding the various p_a (probability of batches being accepted) values, for various P (fraction or % defective) values. It is usually only necessary to locate two points on the curve at ordinates $(1 - \alpha, AQL)$; $(\beta, LTPD)$, as shown at Fig 14.20. The probability values p_a can be derived from the binomial distribution. [Or the poisson distribution if N is very large, $nP \leqslant 5$ and $P \leqslant 0 \cdot 1$. (Note that mean number of defectives in a batch $= nP = z$ in poisson distribution.)] Let us take an example.

Example 14.17

Over a long period of time, the incoming quality of batches of bought-out components delivered to a factory are found to be:

$$AQL = 2\%, LTPD = 5\%$$

A single sampling scheme is to be used for random sampling, reading thus:

'Inspect random samples of 100 from lots of 1 000. If 0 to 3 defects are found in the sample, accept the lot. If more than three defects are found in the sample, reject the lot. Calculate the producer's risk and the consumer's risk.'

Solution

$$AQL = 0.02, LTPD = 0.05$$
$$n = 100, c = 3, N = 1\,000$$

The mean number of defects expected per sample $= nP = 100 \times 0.02 = 2$.

Calculating $(1 - \alpha)$. Using poisson expression with $z = 2$, calculate probability of finding three or less defectives.

This is given by:

$$e^{-z} + z(e^{-z}) + \frac{z^2}{2!}(e^{-z}) + \frac{z^3}{3!}(e^{-z})$$

p(0 def.) p(1 def.) p(2 def.) p(3 def.)

$$= e^{-2} + 2(e^{-2}) + \left[\frac{4}{2 \times 1}(e^{-2})\right] + \left[\frac{8}{3 \times 2 \times 1}(e^{-2})\right]$$
$$= 0.135 + 0.27 + 0.27 + 0.18 = 0.86$$

Therefore p (3 or less defectives) $= 0.86 = 1 - \alpha$

Therefore producer's risk $\alpha = 0.14 = 14\%$

Calculating β

$$\text{Here } nP = 100 \times 0.05 = 5$$

p (3 or less defectives) at $z = 5$ is given by:

$$= e^{-5} + 5(e^{-5}) + \left[\frac{25}{2 \times 1}(e^{-5})\right] + \left[\frac{125}{3 \times 2 \times 1}(e^{-5})\right]$$
$$= 0.007 + 0.035 + 0.088 + 0.147 = 0.277 \text{ say } 0.28$$

Therefore consumer's risk $= \beta = 28\%$.

In practice, a *cumulative poisson probability chart* is used from which the values of 0.86 and 0.28 can be read directly. These charts can be found in any standard work on statistics.

Designing a Single Sampling Scheme

Once the O.C.C. has been established, and the value of N is known, how do we decide upon the values of n and c in order that the amount of inspection shall be at a minimum? These values can be deduced from first principles, but the computation is rather long and tedious. Fortunately, this is not necessary because sampling schemes have been tabulated and published, the Dodge-Romig plans (Reference No. 1 in Further Reading) being based upon the principles outlined in this section. In these plans, β is constant at 10%, and with the other statistical parameters known, n and c can be read directly from the tables. The values derived from the D-R plan will ensure that the average amount of inspection carried out for any scheme will be a minimum.

Cost of Inspection

This will be at a minimum using D-R values. In order to find its value for any particular sampling scheme, we must first compute the *average total inspection* for the scheme. (Remember that as the value of P fluctuates in each sample, so the amount of inspection to be carried out will fluctuate between n and N for each lot received. So we must consider an average value.)

For single sampling:

$$ATI = n + (1 - p_a)(N - n)$$

This will give us the average amount of inspection to be carried out (in terms of numbers of parts inspected) in the long run, for lots of a certain value N.

Example 14.18

An O.C.C. has the following values, $\beta = 0.1$, $a = 0.05$, $LTPD = 0.05$ and $AQL = 0.02$. The incoming lots are 2 000. Calculate the average total inspection for a single sampling scheme.

Solution

From the D-R tables, at $N = 2\,000$ and the statistical values given, $n = 282$ and $C = 9$.

$$ATI = n + (1 - p_a)(N - n)$$
$$\text{and } nP = 282 \times 0.02 = 5.64 = z$$

[From first principles, or alternatively poisson cumulative probability chart, p (9 or less defectives) $= 0.94$, at $z = 5.64$.] $\therefore$ $1 - p_a = 1 - 0.94 = 0.06$.

Therefore $ATI = 282 + (0.06)(1\,718) = 385$.

i.e., On the average 385 parts will be inspected per batch. If the cost of inspection per part is known (i.e., labour cost, overheads, etc.), then the total cost of inspection can quickly be found, this being a minimum for example given.

The same amount of statistical protection can be given if different values of n and c are used for $N = 2\,000$, but the ATI will not then be at a minimum.

Check. For Example 14.18, try C values of 8 and 10 (instead of 9), and calculate the ATI.

At $C = 8$ and $p_a(\beta) = 0.1$; $z = 13$ (from poisson)

Therefore $n = \dfrac{nP}{P} = \dfrac{z}{P} = \dfrac{13}{0.05} = 260$

Then $ATI = 399$.

At $C = 10$ and $p_a(\beta) = 0.1$; $z = 15.5$ (from poisson)

Therefore $n = \dfrac{15\cdot5}{0\cdot05} = 310$

Then $ATI = 395$.

Either of these alternative schemes, while giving required protection according to the O.C.C., will cost more to operate.

14.11 DOUBLE SAMPLING SCHEME

Inspectors often do not like the 'sudden death' result from a single sampling scheme, in which the lot might be rejected on the evidence of a single sample. It is often felt that a second sample should be taken, in order to reinforce the evidence of the first sample, before commencing to inspect the whole batch N.

Double sampling schemes provide for this, and such a scheme is shown diagrammatically at Fig 14.21.

This can be interpreted as:

Inspect a sample n, from lot N.

If less than C_1 defects are found in the sample, accept the lot.

If more than C_2 defects are found in the sample, reject the lot.

If between C_1 and C_2 defects are found, take another sample.

The operation is now treated as a single sampling scheme where $(n_1 + n_2)$ is the sample n, and C_3 is the acceptance value.

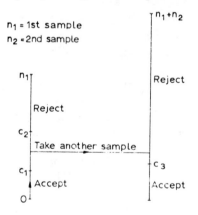

Fig 14.21   Double Sampling Scheme.

n and c values can be taken from the D-R tables for double sampling schemes, and it will be found the ATI is less for double schemes than corresponding single schemes.

For example, comparing two statistically identical schemes, (i) single, and (ii) double, for lots of 2 000; (i) gives $c = 1$, $n = 75$ and ATI is found to be 181. (ii) gives $c_1 = 0$, $c_2 = 4$, $n_1 = 80$ and $n_2 = 165$, and *ATI* is found to be 151.

However, it will be appreciated that administrative costs of operating double schemes are greater than for single schemes.

14.12 MULTIPLE (SEQUENTIAL) SAMPLING SCHEMES

The principle of double schemes can be extended to multiple schemes, in that many small samples are taken of the same value, until a decisive result for the lot is obtained.

A multiple scheme is shown diagrammatically at Fig 14.22.

The nomenclature is similar to that of the other two types discussed. It can be seen that the indecisive zone between C_1 and C_2, C_3 and C_4 etc., may result in the inspector taking sample after sample, until the lot N is inspected, still with no decisive result being obtained. On this ground, such schemes are often criticized as being indecisive. However, it is possible to place a critical limit at some stage which determines whether the lot is finally rejected or accepted.

The advantage of multiple schemes is that the *ATI* value is lower even than for double schemes, but of course such schemes are more difficult to administrate.

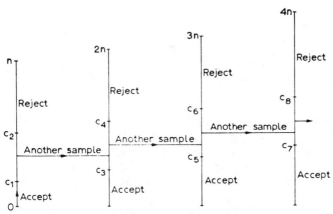

Fig 14.22 Multiple Sampling Scheme.

Sequential Schemes

The principle of multiple sampling can be extended until samples of one only are being taken. This type of sampling is called sequential sampling (being a form of multiple sampling) and can be represented in graphical form, as shown at Fig 14.23.

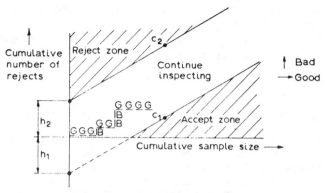

Fig 14.23 Sequential Sampling.

For a given O.C.C. the points h_1, h_2, C_1 and C_2 (at any given cumulative value of n), can be computed. Again, however, sequential schemes have been published, and the Columbia University Tables (Reference No. 2 in Further Reading) can be used, from which these values can be directly extracted.

When the limits for the acceptance and reject zones have been drawn, sampling of random parts takes place. When a *good* or *bad* part is found, it is marked on the sampling graph with a horizontal or vertical line respectively, as shown at Fig 14.24. This line is drawn to a unit value, in accordance with the ordinate scales used on the graph. When these cumulative horizontal (G) or vertical (B) lines, as a result of sampling, eventually cross either zone limit, then the lot is either accepted or rejected. If the cumulative lines stay in the indecisive zone, then another sample of one is taken, and so on with successive samples of one chosen at random, until a decisive result is obtained.

Exercises 14

1. The diameters of nine components are measured and found to be: 5·49, 5·67, 6·17, 3·46, 4·51, 4·11, 6·75, 4·01 and 6·90 mm respectively. From first principles, calculate: [a] the arithmetic mean ($\bar{x}$), [b] the standard deviation (s), and [c] the standard deviation of a distribution of means drawn from the same population (s_n), when size of samples is 4.

(Ans. [a] $\bar{x} = 5·23$ mm, [b] $s = 1·19$ mm, [c] $s_n = 0·595$ mm)

2. The following results in mm are obtained from measuring the lengths of 34 components taken at random from a process.

12·24	12·70	11·80	10·81	9·80	12·65	12·25
11·15	11·52	10·63	11·19	13·15	12·54	10·20
12·10	9·98	9·92	12·80	10·60	11·60	12·65
10·50	13·55	11·10	10·88	12·73	11·46	12·40
10·21	12·45	10·80	10·32	13·02	12·20	

Estimate the values of [a] arithmetic mean $(\bar{x})$, and [b] standard deviation (s).

(Ans. [a] $\bar{x} = 11\cdot61$ mm, [b] $s = 1\cdot11$ mm).

3. If two cards are drawn from a pack, calculate the probability that both are aces.

$$\text{(Ans. } p = \frac{^4C_2}{^{52}C_2} = \frac{1}{221})$$

4. A batch of 50 components contains three defective items. (i) If one part is selected at random, calculate p (it is defective). (ii) If two parts are selected at random calculate p (one of the two is defective). (iii) If three parts are selected at random, calculate
[a] p (one of the three is defective); [b] p (all three are defective).

$$\text{(Ans. (i) } \frac{3}{50} \quad \text{(ii) } \frac{141}{2\,450} \quad \text{(iii) [a] } \frac{3\,243}{58\,800} \quad \text{[b] } \frac{3}{58\,800})$$

5. A random sample of four items is selected from a lot of 1 000 items containing 50 defectives. Calculate (a) the binomial, and (b) the poisson probabilities that the sample will contain exactly 0, 1, 2 and 3 defectives.

(Ans.	p (0 *def.*)	p (1 *def.*)	p (2 *def.*)	p (3 *def.*)
(a)	0·814	0·171	0·013 5	0·000 5
(b)	0·819	0·164	0·016 4	0·001 1)

6. A distribution of weights is normally distributed, having a mean value of 11·504, and a standard deviation of 0·211. Using normal cumulative probability values from a table, calculate the probability of values greater than 11·800 being obtained in the distribution.

(Ans. p (values greater than 11·800) = $8\cdot04\%$)

7. Samples of 200 components taken at random each day, are gauged as a means of checking. The following results are given for the first 24 days, in terms of fraction defective, (P).
0·45, 0·5, 0·45, 0·6, 0·35, 0·55, 0·65, 0·75, 0·75, 0·65, 0·45, 0·55, 0·5, 0·6, 0·35, 0·5, 0·45, 0·45, 0·5, 0·5, 0·6, 0·55, 0·55 and 0·6.
a) From this sample, calculate $\bar{P}$, and the position of the action and warning limits.
b) Draw a P chart, and plot the results of the 24 days sampling upon the chart. Is the process in statistical control for this period of production?

(Ans. (a) $\bar{P} = 0\cdot53$.   U.W.L. = 0·60, L.W.L. = 0·46
U.A.L. = 0·635, L.A.L. = 0·425)

8. 25 samples of 20 components were taken at random from a process, and the number of defectives in each sample were:

2, 2, 1, 0, 2, 1, 4, 4, 5, 2, 1, 2, 1, 0, 1, 3, 3, 0, 2, 1, 2, 0, 0, 5, 0.

Derive the limits, and draw up a control chart for number defective. Plot the above results upon the chart.

$$(\text{Ans. } \bar{m} = 1.8 \quad \begin{array}{l} \text{U.A.L.} = 6.5 \\ \text{U.W.L.} = 4.7.) \end{array}$$

9. Twenty samples of four components per sample were taken from a process, and the length of the parts measured in m. The results are as follows:

Sample No.				
1	1·16	1·25	0·66	0·56
2	0·84	0·82	0·92	0·60
3	0·97	0·94	0·99	0·90
4	1·00	0·94	1·50	1·18
5	0·75	0·97	0·47	0·73
6	0·92	0·60	0·82	1·14
7	1·17	1·00	0·85	0·36
8	0·68	0·93	0·89	1·13
9	1·00	0·91	0·60	0·68
10	0·97	0·87	0·71	0·89
11	0·73	0·66	0·79	0·59
12	0·82	0·77	0·67	0·70
13	0·90	1·25	1·00	0·81
14	0·57	0·62	0·61	0·69
15	0·61	1·02	1·45	0·93
16	0·81	1·00	1·25	0·90
17	0·71	0·94	0·87	0·84
18	0·97	1·06	1·10	0·89
19	1·12	0·73	0·62	0·78
20	0·68	0·61	1·00	1·11

Draw $\bar{x}$ and w charts. Plot the given results upon the charts, and determine whether or not the process is in statistical control.

(Ans. $\bar{x}$ *Chart*   U.A.L. $= 1·17$ L.A.L. $= 0·56$
       $\bar{X} = 0·864$ U.W.L. $= 1·05$ L.W.L. $= 0·67$

       w *Chart*
       $\bar{w} = 0·406$ U.A.L. $= 1·05$ L.A.L. $= 0·04$
                    U.W.L. $= 0·79$ L.W.L. $= 0·12$)

10. If $\bar{w}$ for samples of three is 25 mm, and the specification tolerance is 140 mm, calculate the R.P.I. and indicate its category.
(Ans. *R.P.I.* $= 5·6$)

11. In a single sampling scheme, $n = 71$, $c = 1$, $N = 500$, $AQL = 0·005$, $LTPD = 0·05$.
a) Calculate α and β and draw the O.C.C.
b) Calculate the ATI.
(Ans. (a) $\alpha = 5\%$, $\beta = 13\%$, (b) $ATI = 97$)

12. With the aid of sketches, show the principle of (a) single, (b) double, (c) multiple and sequential sampling. Compare each type of scheme.

Futher Reading

1) Dodge H. F. and Romig H. G. *Sampling Inspection Tables.* Chapman and Hall.

2) *Sampling Inspection.* Columbia University Press.

3) 'Sampling Procedures and Tables for Inspection by Attributes.' *Defence Specification* 131 HMSO.

4) 'Application of Statistical Methods to Industrial Standardization and Quality Control.' BS 600.

5) 'Fraction Defective Charts for Quality Control.' BS 1313.

6) 'Control Chart Technique.' BS 2564.

7) Moroney M. J. *Facts from Figures.* Penguin Books.

8) Grant E. L. *Statistical Quality Control.* McGraw-Hill Book Co. Inc.

9) Jenney B. W. and Newton D. W. 'Statistics and Quality Control.' *Journal of Institute of Production Engineers.* March 1969.

10) Roebuck E. 'Machine and Process Capability Assessment and Process Control'. *Journal of the Institute of Production Engineers*, May and June 1973.

11) Croasdale H. 'Inspection by Sampling'. *Journal of the Institute of Production Engineers*, September 1979.

Index